Santa Monica Public Library

I SMP 00 2100281 L

SANTA MONICA PUBLIC L   O9-CFS-984
                            APR   2009

Hijacking Sustainability

Hijacking Sustainability

Adrian Parr

The MIT Press
Cambridge, Massachusetts
London, England

© 2009 Massachusetts Institute of Technology

All rights reserved. No part of this book may be reproduced in any form by any electronic or mechanical means (including photocopying, recording, or information storage and retrieval) without permission in writing from the publisher.

MIT Press books may be purchased at special quantity discounts for business or sales promotional use. For information, please email special_sales@mitpress.mit.edu or write to Special Sales Department, The MIT Press, 55 Hayward Street, Cambridge, MA 02142.

This book was set in Stone Sans and Stone Serif by the MIT Press. Printed and bound in the United States of America.

Printed on recycled paper.

Library of Congress Cataloging-in-Publication Data

Parr, Adrian.
Hijacking sustainability / Adrian Parr.
  p. cm.
Includes bibliographical references and index.
ISBN 978-0-262-01306-2 (hbk. : alk. paper)
1. Sustainable development—Social aspects. 2. Environmentalism—Social aspects.
3. Social change. I. Title.
HC79.E5P353   2009
338.9'27—dc22

2008029420

10 9 8 7 6 5 4 3 2 1

To Lucien and Shoshana

Contents

Acknowledgments

In writing this book many people have generously and kindly given up their time to read over draft chapters and provide comments. I would particularly like to thank Ronald Bogue, Ian Buchanan, Felicity Colman, Mike Parr, Paul Patton, and Kenneth Surin. I have been lucky to have their feedback, especially considering they are the people whose own work has had a profound impact on me over the years. Versions of chapters 1 and 7 were published as part of the "Seeking the City" 96th ACSA Annual Meeting Conference Proceedings. In addition, I am most grateful for the graduates in the School of Architecture and Interior Design program at the University of Cincinnati who provided me with intellectual stimulation on a weekly basis. Their questions and lively discussions helped shape the final manuscript. Finally, I need to thank Michael Zaretsky, a talented architect committed to the principles of sustainable design, for providing me with endless resources, expert suggestions, detailed discussion, steady editing, unwavering support, and inspiration. I owe the completion of this book to him.

Acronyms

AMA	American Marketing Association
BAN	Basel Action Network
BFA	Browning Ferris Industries
CFC	Chlorofluorocarbon
DoD	Department of Defense
EMA	Environmental Media Association
ENR	Environment and National Resources
EPA	Environmental Protection Agency
FHA	Federal Housing Administration
FSC	Forest Stewardship Council
ICC	Inuit Circumpolar Council
IDP	Internally Displaced Persons
IPCC	Intergovernmental Panel on Climate Change
IUCN	International Union for the Conservation of Nature and Natural Resources
LEED	Leadership in Energy and Environmental Design
MDG	United Nations Millennium Development Goals
NGO	Nongovernmental Organization
NRDC	National Resource Defense Council
OEOB	Old Executive Office Building
OLPC	One Laptop Per Child
OSHA	Occupational Safety and Health Administration
SEI	Strategic Environment Initiative

SIPRI   Stockholm International Peace Research Institute

SRI   Socially and Environmentally Responsible Investment

SUSG   Sustainable Use Specialist Group

UNHCR   United Nations High Commission for Refugees

VOC   Volatile Organic Compounds

WCED   World Commission on Environment and Development

WMI   Waste Management Inc.

Introduction

Sustainability. Gone are the days when the word conjured up images of unapologetic veganism, dreadlocks, and mud-brick homes. From ecohippie to ecohip, *sustainability* is the new buzzword on the lips of many Americans. The corporate sector is going green, Hollywood is taking up the cause with a bang, cities are being ranked according to how sustainable they are, and popular media are increasingly shifting their attention onto the problem of how the United States can change color. Why have these disparate lines of cultural production begun to convert to the green cause? Some might say natural disasters such as Hurricane Katrina, which ravaged New Orleans on August 29, 2005, have pushed the issue of global climate change into the frontal lobe of the popular imaginary, engaging a deepened sense of environmental responsibility where previously there had been none (or very little). Others salivate at the prospect of new markets and growing profits, rubbing their hands together in anticipation of the loud cha-ching cha-ching that their new investments in green technologies ring forth.

So, amidst all the heated fervor what exactly does sustainability refer to? The most commonly upheld definition comes from *Our Common Future*, the 1987 report on the state of global natural resources and the human environment compiled by the World Commission on Environment and Development (WCED).[1] With a clear call to global cooperation the report, developed under the guidance of the former first woman prime minister of Norway and Chair of the WCED Gro Harlem Brundtland, explains that we need to combine social, economic, and political concerns if we are to successfully move toward a more sustainable future. In it sustainability is understood as development that meets the needs of today without compromising the needs of future generations. The report clarifies two very important definitions of what constitutes the environment and development: "the 'environment' is where we all live; and 'development' is what we all do in attempting to improve our lot within that abode. The two are inseparable."[2]

With the primary objective of inaugurating a new era of economic growth that demands a return to multilateralism, the Brundtland Report insists that "people can build a future that is more prosperous, more just, and more secure."[3] Emphasis is given to a new international economic structure that fosters long-term cooperation, one that assigns an important developmental role to multinational companies and multilateral financial institutions especially in respect to initiating sustainable development initiatives in developing countries. Yet how do we reconcile the great divide that emerges when international organizations, for all their supposedly good intentions, bulldozer local specificity in the name of international aid and large-scale intervention?

The macro perspective of internationally coordinated sustainable-development initiatives can be held in stark contrast to grassroots initiatives operating at the local level (what are commonly referred to as social and environmental justice groups, which use a bottom-up or microeconomic approach). Starting out with the disparities between groups with access to environmental goods—such as unpolluted and sanitary living conditions—and those who carry too many environmental burdens—such as communities in proximity to landfills and industrial waste—the thousands of local organizations that constitute the sustainability movement seek to eliminate the power structures underpinning the disproportionate burden of social and environmental ills underprivileged groups carry. Like the appeal to the "rights of the human family to a healthy and productive environment" made in the Brundtland Report, the sustainability movement also takes a rights-based approach. That is, the movement seeks justice for the underprivileged, including the right of the environment not to be destroyed. However, unlike recommendations made in the Brundtland Report, the solution proffered is not a multilateral global approach to the problem of sustainable development; instead these groups are often more interested in initiating and supporting local programs.

The irony, however, is that grassroots organizations still need to squeeze local specificity into a manageable and general rubric before the needs of disenfranchised groups can be represented in the political arena. In reality, power structures are challenged and critiqued only after representation is reintroduced into the political vocabulary, and its effectiveness depends upon the collaboration of large institutions (governmental and/or international) as much as it does the efforts and commitments of local actors. For these reasons, problems of representation and agency are what frame my discussion of sustainability culture throughout this book. I propose that the politics of sustainability culture arises in the way culture engages with problems of environmental exploitation and social injustices with a view to supporting and activating a sense of agency for disenfranchised individuals and groups.

As the public's enthusiasm for sustainable ways of life, environmental stewardship, and social equality grows, popular culture is rapidly becoming the predominant arena where the meaning and value of sustainability is contested, produced, and exercised. To state the obvious, this is because sustainability culture is a social practice. It is an instrument of knowledge formation; it is how a local context is narrated; it engages new and emerging social values and the energies driving these in dialogue with more traditional values and conventions, along with the habits and stereotypes underscoring these.

As the first half of this book demonstrates, the power of sustainability culture is not one sided, it is an affective encounter simultaneously crisscrossing a multiplicity of trajectories. As the popularity for green commodities grows the public's enthusiasm for the principles of sustainability increases. This situation has produced a rising interest in the ethics of business practice and ushered in a new kind of shareholder activism. It involves movie and sports stars putting their influence and power throughout the popular imaginary to work for activist causes. Even former U.S. presidents jumped on the bandwagon, appropriating the symbolic power of the White House to showcase their commitment to environmental stewardship. These are all examples of sustainability culture at work.

Whether it is the hard-core activist living in treetops in an effort to save the wilderness, or the right-wing conservative vigorously disputing the scientific accuracy of the theory of global climate change, or architect William McDonough and chemist Michael Braungart tempting industry into a marriage of economic convenience with the "greenies," as Fredric Jameson might say all these positions on sustainability, "whether apologia or stigmatization—[are] also at one and the same time, and *necessarily*, an implicitly or explicitly political stance on the nature of multinational capitalism today."[4] This book will add into the mix historian Andrew J. Bacevich's thesis on the seductive power of militarism, proposing the deepest challenge sustainability culture faces is the increasing militarization of life, which is bound up with the logic of late capitalism (global markets, multinational corporate activity, outsourcing of labor, and the important role the media plays in promoting consumer culture), and in turn argue it is the poor who largely bear the brunt of both.[5]

Although action is undoubtedly necessary, in order to be truly long-term the significance of locating sustainability within its own concrete historical condition of global capitalism and increasing militarism is of paramount importance. Indeed, sustainability culture serves as an engine for social change, one that is defined by historical breaks and continuity. I argue the historical contingency of sustainability culture comes from the sociopolitical dilemmas culture works with, expanding and intensifying social life so as to reinvent how we live. Put differently, sustainability culture is how societies designate the specificity of their historical condition in material form and, as such, the concern is not so much with ends (the utilitarian focus on meeting and maximizing needs) as it is how sustainability becomes a political attitude of the multitude. In other words, sustainability culture is inherent to the logic of late capitalism and, therefore, the productive force of that culture comes from how it generates economic value (as McDonough and Braungart assert) as well as political currency.

The focus given to new technologies and the economic benefits of these among those involved in sustainable design, development, and practice could be seen as a vestige of the Enlightenment's value in human reason and its overall focus on market economics; moreover, this value is at the heart and soul of multinational corporate culture and liberal politics. This is where the ecobranding efforts of the corporate sector offer an intriguing case study. Chapter 1 examines the ecobranding tactics used by corporations such as British Petroleum (BP) and Wal-Mart as they ride the wave of sustainability culture, expanding their market base as they aggressively promote a new socially and environmentally responsible corporate image. Simply put, both companies try to tap into the rising popularity of socially responsible consumption to maximize their profits. Yet, how they link the social value of caring for the environment with a supposedly renewed image of corporate behavior amounts to nothing more than a modification of what architect Rem Koolhaas calls *junkspace*.[6] For BP and Wal-Mart, their new ecobrands aim to offset the perception of corporate excess by promoting an image of corporate responsibility that relies on the idea that a corporation can use its power to introduce a sense of sustainable consumption into the shopping equation. I remain unconvinced by the corporate beast reincarnating itself as man's best friend, for along with the dog we also inherit the fleas.

Like corporations, Hollywood is also tuning in to the popularization of the sustainability movement. What Hollywood shares with the corporate sector is the manner in which it relies upon a system of reification (in this case, labor is understood to be a commodity that can be bought and sold for profit) and its dependence upon the abstract surplus value (the difference between what the labor to produce a commodity costs and what commodities are actually sold for) such a system puts into play. In what way, if at all, can the labor power of Hollywood not only enter the trope of sustainability but also participate in its production? To explore this question, chapter 2 looks to the appearance of actor Leonardo DiCaprio and Knut on the cover of *Vanity Fair: The Green Issue* (April 2007)—the image of the baby polar bear, who was born into captivity, was much publicized— placing Hollywood activism within a broader narrative that includes the making of the documentary *An Inconvenient Truth*, the greening of the Oscars, and President Bush Junior's poor environmental record and overall reluctance to list polar bears under the Endangered Species Act.

The benefit of looking to cultural production in the context of sustainable development is that culture is not simply ideological. That is, culture not only promotes social awareness of environmental issues; as a practice it has the power to also put sustainable living to work. And it is this pragmatic side of cultural production where it becomes a dynamic system of social, economic, and political activity. It is one that can potentially improve the health and well-being of a community as it promotes principles of equality, stewardship, compassion, renewal and sustenance. This is understood in the following manner. First, future well-being is included in the decision-making processes of the present. The present time is fine-tuned with an awareness of and sensibility for the future without being patronizing. In other words, the present is humbled by the future. Second, communal spaces are shaped and informed by difference, embracing other ways of life for the radical alternatives they pose to what is currently on offer. Connections to the environment and historical circumstances are made so as to foster fresh social organizations that present new opportunities for living. This comes from maximizing local conditions, all the while remaining alert to the impact such connections produce. As discussed in chapter 3 through a comparison of the ecovillage and the gated community, this is a mode of sustainable production understood as the creation of self-organizing communities—which is not to be confused with being independent of context. Indeed, both advance a concept of autonomous living, yet in the context of the ecovillage autonomy becomes a precondition for an affective encounter with the world, whereas for the gated community autonomy quite simply is an extension of the growing culture of militarism throughout mainstream U.S. society.

The affective power of sustainability culture, however, is not always affirmative. It can take on a more conservative and reactive flavor when used to discipline everyday life. This is particularly evident in the numerous greening and de-greening initiatives on the White House carried out by various presidents over the years. Reviving the now-somewhat-unpopular conjunction of politics and aesthetics, chapter 4 examines how issues of sustainable design also constitute an exercise of power. I suggest that the politics of sustainability culture is predicated not so much upon ideological struggles but rather on how a space-time sensorium is organized and constructed. Moreover, sustainability culture is depoliticized when this space-time sensorium becomes a function and exercise of state power.

The more the power of sustainability culture is appropriated by the mechanisms of State and corporate culture, the more it camouflages the darker underbelly of both—militarism and capitalism. In chapter 5 I argue that the policy to green the U.S. military is questionable at best. Given that the function of the military is to conduct war, it is crucial we do not subsume the values that the military propounds with those of civil society. I am clearly biased against the military as I openly claim that the values of military and civil society are anathema to one another, which should not be taken to mean the military is unnecessary; rather, my point is that the attempt to green the military simply enables it to *disguise* the violence it perpetrates against the values of civil society. I propose that the policy to produce a series of sustainability goals for the military is nothing other than what Jacques Rancière describes as the hatred of democracy.

At this point I attend to some of the key challenges facing sustainability culture—trash, disaster relief, slums, and poverty. The underlying current running throughout this discussion is the importance of adequately addressing the needs of the most disenfranchised members of the global community: the poor. Although the tendency is to reach out to the world's extreme poor in the developing world, it is also important to be mindful of the poverty that knocks on our own backdoor. As Slavoj Žižek once cogently remarked:

Every exclusive focus on the First World topics of late-capitalist alienation and commodification, of ecological crisis, of the new racisms and intolerances, and so on, cannot but appear cynical in the face of raw Third World poverty, hunger, and violence; on the other hand, attempts to dismiss First World problems as trivial in comparison with "real" permanent Third World catastrophes are no less a fake—focusing on the "real problems" of the Third World is the ultimate form of escapism, of avoiding confrontation with the antagonisms of one's own society.[7]

In chapter 6 I not only discuss the international trade in e-waste that pollutes the environment and the bodies of the developing world, I also look at the unequal distribution of waste as it clogs the aorta of the U.S. Rustbelt, which is turning the area into an abject landscape whose soil is being filled to the brim with the refuse of wealthier states.

The ethical force of sustainability culture comes from the way in which it manages the inherited power relations of a specific social fabric with sensitivity, all the while scrupulously examining all the discontinuous elements that produce a given community—political, economic, psychological, cultural, and sexual. Culture has the power to construct conditions conducive for social discourse to take place. People can also assert their sense of agency by resisting hegemonic systems of signification that constitute their oppression. In this way, the very pragmatic focus of the design disciplines are particularly well positioned to make a difference.

The Intergovernmental Panel on Climate Change (IPCC) estimates that by 2050 there could be as many as 150 million environmental refugees worldwide.[8] With this in mind, I recognize the important role that the design field will have in helping alleviate some of the problems that arise as large populations of people become displaced. The combination of technical knowledge, practical focus, and creative experimentation indicative of the design field means it is able to directly alleviate some of the debilitating effects natural disasters wreak on the lives of individuals, families, and entire communities. As with the previous example of trash, the communities hit hardest by natural disasters are more often than not the poor. For designers the question now becomes one of not only how to stitch back together social networks that help communities thrive, but also to design in a manner that fosters a sense of agency once more; only then can design interventions be truly sustainable.

Recognizing that a sustainable design is not something that is performed upon a subject, the politics of sustainability culture is, as Judith Butler might argue, a matter of how a subject is brought into being and then how the subject reiterates or contests the "discursive conditions of its own emergence."[9] Now more than ever, as people converge upon the cities of the world and a new urban order emerges in the form of slums, urban designers and planners have a mammoth task ahead of them. I show how governments and aid organizations are starting to use the "informal" economies (a term I question) typical of urban squatter communities ideologically. That is, the status of "informality" that is used to define slum dwellers positions them negatively (in opposition to the formal city/ government and dominant ways of life). As such, it is unsurprising that urban designers and planners aspire to integrate slums into the broader urban fabric. As chapter 8 outlines, however, the model of urban integration is premised upon the distinction between an unintegrated and an integrated urban form, and this distinction produces a particular way of conceiving and knowing the slum dweller's body—one that is defined in opposition to the law and order of the formal city.

Commencing with the position that the city is a material fact, chapter 8 demonstrates how the discourse of urban design and production is developed in accordance with how such facts are used and interpreted. Put differently, urbanism is a process of signification and the matter it designs, frames, and regulates cannot be dissociated from the values and norms it assumes and uses to interpret the city. Studying the Favela-Bairro program in Rio de Janeiro, which aimed to integrate the favelas into the broader fabric of the city, the design began with the presupposition that the slums are other-than-the-city-proper. Ironically, the model of urban integration is complicit in producing two different urban identities, one of which holds a dominant position in respect to the other. Hence, all integration really did in the context of Rio's favelas was open the slums up to the free market and boost the militarization of space as police entered the area and concomitantly fueled the drug wars.

What the Favela-Bairro program of urban integration and development also brings into relief is the manner in which the finger of blame tends to be pointed in the direction of poor communities, declaring their way of life as a barrier to sustainable living and development. However, I also show in chapter 9 that the real challenge sustainability culture faces is not so much the poor themselves but rather the negative perception of the poor as not contributing to the formal city and/or economy. For this reason, the aim to integrate squatter communities misses one fundamental issue: poverty is not simply a problem of property rights, infrastructure, and services. Although microcredit has proven to make enormous differences in the lives of the poor, the success of these programs has less to do with a profit-driven conception of free market economics than with the redistribution of resources and access for the poor to the surplus value of their labor along with training and educational initiatives that loosen the grip of patriarchal structures of violence and oppression. As I outline, this also suggests that multinational corporate economics are not the only source of wealth; there are other avenues for wealth creation that involve a more productive understanding of matter, labor, and energy.

Our current historical condition is one of global climate change, multinational and financialized capitalism, increased religious fundamentalism, and rising militarism—all of which cannot be disentangled from other structural disparities defining the social and economic relations of the developing and developed world. This book examines the new culture of sustainability and how these hegemonic relations are challenged in an effort to revitalize collective life. The challenge can be likened to a utopian impulse operating throughout popular culture. Although utopianism is often associated with the ideology of a particular political position, promising the masses a perfect—albeit impossible—social order, there is another way of understanding utopia. That is a utopia is less driven by content (an ideal defined in opposition to reality) and more by form (the role of utopia in prompting us to think differently about our current situation). The utopian impulse of culture arises when culture registers our current conditions (such as global climate change, militarism, and capitalism) but then transforms these in the process. Hence, the thesis of this book is that ideology in and of itself is not "bad"; it all depends upon how it functions—does it instigate a fixed identity, or does it generate difference? Politically, the question addresses the problem of representation. Culturally, the question broaches the conditions of the Real—the raw state of nature that language cannot represent—so that in order to retain its political bite cultural production necessarily remains resilient to representation. The latter, therefore, introduces the failure of culture to fully announce and articulate utopia, while at the same time the utopian mode of its production relies upon this failure.

As is discussed through the use of case studies in the first half of this book, culture can promote a sense of dignity and care for the environment in ways that institutions, bureaucracies, and governments cannot. This is because culture is an especially utopian praxis, but not in the sense that it creates an imaginary ideal; rather it exposes, develops, questions, and abstracts the potential and concrete specificity of our present circumstances, all with a look to creating a future that is critically different from what currently is and has been. And as the second half of this book attests, the concept of "criticality" appears in an effort to historicize and evaluate the dominant presupposition of integration and development framing the discourse of sustainability in favor of notions of difference and renewal.

A few years into the twenty-first century, why has sustainability really begun to take hold of the U.S. imagination? It is no accident that the enthusiasm for all things green gathered momentum at a time when the country started to become demoralized and disillusioned with the failing war in Iraq and the political mess-ups of the fumbling administration of Bush Junior. That is, sustainability offers an alternative narrative to the one of never-ending militarism; it is one that promises to clear the skies with a renewed sense of optimism for a future different from the present and past, at a time in history when the formal political arena offeres only fear tactics in place of promise and vision. It is this utopian dimension of reaching toward an alternative lifestyle that this book narrates. The works and modes of cultural production presented throughout the pages that follow vary widely, and yet they all speak to a different way of understanding sustainable practice. Despite these differences they all share a similar concern with putting the limits and constraints that arise out of social, economic, and environmental hardship to work in productive and creative ways.

Educated Consumer. Photograph by Michael Zaretsky, 2007.

I The Popularization of Sustainability Culture

1 The Greening of Junkspace

One says "pleasure in a thing": but in reality it is pleasure in oneself by means of a thing.
—Friedrich Nietzsche, *Human, All Too Human*[1]

At the turn of the twenty-first century, sustainability has rapidly become the buzzword on everyone's lips. In the United States it is not ideology that is turning sustainability into a cultural hegemonic: it is a socially and environmentally conscious multitude whose investment and consumption patterns are prompting multinational corporations such as Wal-Mart and BP to develop a new image of corporate social responsibility. The rise of socially (and environmentally) responsible investment (SRI) has prompted corporations to become more accountable and transparent. At the same time, the convergence of popular culture and the sustainability movement has provided the corporate world with new opportunities to resituate their products and services within the competitive global market. Ironically, ecobranding—this new strain of branding committed to showcasing a company's social and environmental responsibility—has been empowered by the emerging culture of activist investment. However, what are the effects of this marriage between commerce and sustainability culture? Does ecobranding produce passive landscapes of consumption or active and affective landscapes conducive to sustainable ways of life?

As social and environmental injustices present themselves, there has been a simultaneous cultural shift toward social and environmental activism. This phenomenon, which I describe as sustainability culture, is an affective state. That is, it is associated with shifts in the state of a body's feeling of power.[2] The feeling of power, however, too often is confused with the power to act as compared with a more affective notion that involves a power to affect and be affected. In other words, instead of asking what power is and where it comes from, we need to ask how it is practiced. A case in point would be the strengthening culture of SRI. These investors align their money with certain ethical positions they hold, such as not financing companies that support military regimes or ones that use child labor. The feeling of power produced in the activist investor simultaneously empowers the corporate sector to change. In turn, these changes affect the commodity market and character of popular culture. That said, just because sustainability culture can change the way the corporate world does business, assuming that this equates with corporate transparency and accountability would be a mistake. As the theory of greenwashing brings to our attention, activist investment does not necessarily change the ills of the global marketplace; corporations also use the affective power of sustainability culture to camouflage otherwise unsustainable business practices. Corporate Watch calls this technique *greenwashing*, "the phenomenon of socially and environmentally destructive corporations attempting to preserve and expand their markets or power by posing as friends of the earth."[3] Greenwashing aspires to change the public's negative perception of a corporation by promoting a new sustainable corporate image. However, the public face of the company is asymmetrically aligned with the way it conducts business.

A case in point would be the multinational oil company BP, originally British Petroleum and later BP Amoco after 1999. BP set out to redefine itself as an environmental company in 1997 when it withdrew support from the Global Climate Coalition, an industry organization established in opposition to the 1997 Kyoto Protocol and that had fought hard against regulating greenhouse gas emissions. The often-cited reason given by the BP chairman for the break was, "the time to consider the policy dimensions of climate change is not when the link between greenhouse gases and climate change is conclusively proven, but when the possibility cannot be discounted and is taken seriously by the society of which we are part. We in BP have reached that point."[4] BP reentered the market with a fresh logo, a name change, and an aggressive marketing campaign under the guidance of renowned advertising agency Ogilvy and Mather. In 2000 the connection between environmentalism and BP was further reinforced with the slogan "Beyond Petroleum," a catchy way to repeat the old association while producing it differently. The slogan quickly helped situate BP in the object space of environmental brands such as the Rainforest Action Network, whose slogan for its alternative energy campaign was "Beyond Oil." As is discussed later in this chapter, this strategy is common for those involved in realist branding initiatives—the equity of the brand is the result of finding its own niche within object space alongside other similar products that are used to legitimate the credibility of the brand and from which the new brand also differentiates itself. So how does BP's social and environmental record align with the socially and environmentally responsible image that its new logo touts?

The story reveals itself as we work backward in time. In July 2007, the *Chicago Tribune* reported that the Environmental Protection Agency (EPA) and the state of Indiana had given BP an exemption from the provisions outlined in the Clean Water Act (1977), allowing the company to dump approximately 4,925 pounds of sludge and 1,584 pounds of ammonia into Lake Michigan.[5] Before that, there was the lawsuit Colombian farmers waged against the company in 2006 after they had been forcibly removed from their farms to make way for a 450-mile BP pipeline. The farmers had been intimidated by Columbian paramilitary groups to relocate to nearby towns, where they lived in slum conditions. After taking their case to the High Court in London, where BP faced a charge of human-rights abuses, the two sides came to a mutual agreement prior to the case going to trial. The company not only avoided a very public and lengthy trial, but it also did not have to admit liability.

A leakage from a corroded Alaskan pipeline on March 2, 2006 resulted in BP indefinitely closing down its Prudhoe Bay operation in August 2006. This negatively impacted upon 8 percent of the U.S. domestically produced oil supply. The Alaskan oil spill of approximately 270,000 gallons of crude oil—the largest spill in the history of the North Slope—resulted in a U.S. grand jury issuing a subpoena to the company in June of 2006.[6] Slightly farther back in time, on March 23, 2005, their Texas refinery exploded, killing fifteen people and injuring more than 170 others; BP was fined $21,361,500 by the U.S. Department of Labor's Occupational Safety and Health Administration (OSHA) for violating safety regulations, which resulted in the explosion.[7] After a thorough investigation took place, the verdict of the U.S. congressional committee was that cost cutting "was having a troubling effect on the company's ability to design and maintain a sound corrosion control program."[8] Furthermore the committee reported that the problems leading to the Alaska oil spills were strikingly similar to those that created the Houston refinery explosion. After the Texas refinery incident, outraged shareholders demanded that BP directors have their bonuses linked to how well the company performs on safety and environmental issues, a clear case of SRI at work.

BP is a recent example of SRI in the United States, but in the past SRI often began with religious groups and indigenous cultures marrying their core beliefs and traditional ways of life to their economic practices. Namely, it was common practice amongst the Quakers and Methodists to invest in ways that did not promote the slave trade. Established in 1928 as primarily a Christian investment fund, Pioneer Fund screened its portfolio of "sin stocks." To this day there are no alcohol, tobacco, or gambling industries in the fund's portfolio. Then, the U.S. civil rights, feminist, and environmental and peace movements of the 1960s raised the public's awareness of corporate responsibility on social and environmental justice issues. By the late 60s and early 70s this interest in SRI became more institutionalized, seeing the establishment of organizations such as the Investor Responsibility Research Center and the Council of Economic Priorities. Also around this time the first contemporary form of socially responsible mutual funds came into being with the Pax World Fund (1971) and the Dreyfus Third Century Fund (1972), both of which avoided investing in companies with shoddy social and environmental records, nuclear power stocks, and military defense contractors.

However, it was not until the anti-Apartheid campaign of the 1980s that the SRI movement really gathered steam in the United States. At this time "social investors and institutions divested their portfolios of companies doing business in South Africa as a protest against the regime's system of racial inequality or led resolutions with companies in operations there."[9] Continuing on from there, Chernobyl and the Exxon Valdez oil spill provided flashpoints for investors concerned with environmental corporate responsibility.[10] In this way, the anti-Apartheid campaign and a series of high-profile environmental disasters produced a break in the socioeconomic arena, one that redistributed social energies so that they found investment in a different mode of economic practice. The result was that the power of the shareholder started to intensify and shape the economic landscape from the bottom up. The anti-Apartheid investment campaign changed the nature of investment. Drawing on the energies and affects that connect with broader collective struggles, SRI redefined the landscape of investment as it energized the voice of the investor, and in the process the culture of the corporation started to enunciate itself differently. As SRI gathered momentum it generated different ways of entering into the market of investment as well as constructing alternative ways of exiting it.

Such socially responsible investors follow "an investment process that considers the social and environmental consequences of investment, both positive and negative, within the context of rigorous financial analysis."[11] This group of investors consists of individuals, universities, businesses, corporations, hospitals, religious and nonprofit organizations, and even pension funds. In the decade prior to 2005 the association between economic life and social and environmentalist culture strengthened, with $1 in every $10 of professional management in 2005 coming from SRI. In 1995, $639 billion—totaling 9 percent of the $7 trillion in total assets under professional management—were managed using SRI strategies. Since then it has increased at an average annual rate of 26 percent; by 2005 SRI reached $2.3 trillion in total assets. From 1995 to 2005 in the United States SRI grew more than 258 percent as compared to other assets under professional management, which grew 249 percent.[12] These figures certainly reinforce the idea behind the Next Industrial Revolution, with environmental and social justice policies being strengthened as they enter into a marriage with the values of big business.[13] Later in this book I challenge this claim more directly, but suffice it say for the moment to briefly add the following caveat: I share the one rule George Monbiot works by, which is to "trust no one who has something to sell."[14]

The three principal SRI strategies include screening, shareholder advocacy, and community investing. The first strategy uses social and environmental criteria to evaluate investment portfolios and mutual funds, much like the previously cited example of the Pax World Fund. In the case of the second strategy, shareholders use their power as owners of U.S. corporations to shape the social and environmental performance of a company. To return to the case of BP, in the wake of the Texas explosion the Local Authority Pension Fund Forum (it owns 1.2 percent of the group's shares) placed BP chair Peter Sutherland under pressure to address how the pay of senior executives could be linked to nonfinancial issues such as social and environmental performance factors.[15] This is a perfect example of shareholder advocacy. The third strategy, community investing, "directs capital from investors and lenders to communities that are underserved by traditional financial services."[16] Combined, they create a political encounter as both the power of the investor and that of the corporation are enhanced in tandem.

The end result of activist investment is a protest portfolio involving both reactive and active uses of power.[17] Activist investment is reactive insofar as it involves a negative screening technique, withdrawing money from companies identified as having a poor social and environmental record. As an active power it seeks out and invests in companies that are enthusiastically involved in promoting the practices of good corporate citizenship. Both modes of investment draw on the notion of power as affective. The idea being, just as the actions of a corporation produce sensations and affects in us and throughout the social field, so too do our own economic practices.

Because it operates outside of the structure of domination, SRI is affective. It constitutes a different exercise of power, which emerges out of a dynamic of agitation as opposed to an ideological struggle between two distinct positions (activist versus capitalist, or the greenwashing thesis that claims corporations ideologically manipulate the public's perception of them). In other words, the investor's own sense of power as a politically engaged citizen and the economic power of the corporate world mutually reinforce each other.

It is evident that the investors' sense of power comes from the feeling that they have the power to not only change the way corporations do business, but also that on a deeper level they increase the power of the corporation and hence their own feelings of power are awakened. In short, the trend toward SRI demonstrates that by contributing to the power of capitalism investors' feelings of power are amplified. As each is affected by and affects the other, their capacities to act and be affected also change, producing a dynamic of creative change. However, one can never tell in advance where this path of transformation and mutation, formed out of the unlikely connection between multinational corporate life and sustainability culture, will go. Will it produce a new organization of a socioeconomic territory, or will it return to the recognizable landscapes of junkspace? More recent trends seem to suggest the latter: as sustainability culture goes corporate, it is turning into another branding strategy.

The changing face of the Hummer is a classic example of what happens when sustainability culture and capitalism join forces. In the aftermath of the 9/11 terrorist attacks a large military vehicle was domesticized and renamed *Hummer*. Acquired by General Motors in December 1999, the Hummer flexed its fuel-inefficient muscles in defiance of the rest of world, as the U.S. and its allies went to war in the Middle East in search of Osama Bin Laden and later Saddam Hussein. Sales almost doubled between July and November of 2002, while at this time the figures for SRI in the U.S. dropped for the first time in ten years (the social feeling of power was now invested in militarism).[18] In 2001, SRI figures were at $2.32 trillion but they fell to $2.16 trillion in 2003.[19] As these numbers improved (by 2005 SRI increased to $2.29 trillion) this translated into an overall shift in the pattern of commodity culture. Entering the United States on the heels of the Hummer, which gets 13 miles to the gallon, sales figures for the Toyota Hybrid Prius, which in comparison gets 46 miles to the gallon, increased 82 percent from January 2003 to January 2004. According to Autodata, by November 2003 twice as many Prius's were sold as compared to the H1 Hummer.

On the whole, hybrids continued to enjoy robust sales, with *Green Car Congress* reporting a 43.8 percent rise in sales between 2005 and 2006 alone.[20] In response to these dramatic shifts in the culture of investment and buying, General Motors developed a new Hummer, which was showcased in the Los Angeles Design Challenge. The concept of the Hummer O_2 is:

a fuel-cell powered vehicle with a phototropic body shell that produces oxygen (O_2) even while parked. The concept features algae-filled body panels that consume atmospheric CO_2 and produce oxygen that is released back into the environment. The O_2's construction specifies the use of 100 percent post-consumer materials like aluminum for the frame and VOC-free finishes.[21]

What we have here is a fascinating crossbreed formed out of an unlikely marriage between ecobranding and militarism. The result was an ironic paradox in terms: a "green Hummer." As General Motors Chief Executive Officer, Rick Wagoner explained:

General Motors is committed to sound corporate citizenship in all aspects of our business. Above all, we know that maintaining a strong company will help ensure our continued commitment to the communities in which we live and work, and to the social interests we have identified as important to our business and our stakeholders.[22]

In other words, the greening of the Hummer was part of a new wave of corporate branding that targets not only the investor but also the consumer.

So, what is a brand? Branding can apply to products such as the Hummer; famous people including former basketball player Michael Jordan; places like Champagne in France and New York City, aka "The Big Apple"; stores such as Wal-Mart or Harrods in London; services including United Parcel Service, better known as UPS; and companies such as BP. More succinctly, the American Marketing Association (AMA) defines a brand as the "name, term, design, symbol, or any other feature that identifies one seller's good or service as distinct from those of other sellers. The legal term for brand is trademark. A brand may identify one item, a family of items, or all items of that seller. If used for the firm as a whole, the preferred term is trade name."[23] As I now briefly describe, there are predominantly two branding positions: idealist and realist.

The idealist maintains that the value of a brand is grounded in consumer perception, such as the psychology and emotions of the individual consumer. Idealists, like Theodore Levitt, hold that the value of a product is external to the product itself. Branding strategies that take this position as their point of departure, set out to influence consumer perception of the brand.[24] Evidently, the usefulness of the product is not as important as the image or associations that the product conjures up in a person's mind (social status, style and so on). Carrying on from brand idealism, the efforts of integrative branding are directed toward strengthening social networks as brand loyalty and a sustainable brand image is built. In this context, the brand carries a social function, setting out to create a long lasting bond with the customer by promoting a set of values and attitudes that it shares with its customers. In short, the public image of the company is supported by the activities of the company in the public sector. Integrated branding sets out to create a consistent customer experience of a company, whether this is communicated in the story, the media, or in the store itself. The idea is that social capital is experienced differently from how goods and services are. According to this view, the clarity and trustworthiness of a brand is crucial to the brand's success on the market. Following on from here, integrative branding tries to engage the emotional and physical experiences of consumers in an effort to create a *"unique and compelling* customer experience," as branding strategists Joseph LePla, Susan Davis, and Lynne Parker explain.[25] What is important are not just the values, principles, and stated mission of the company, or even its identity and the associations these form in the customer's perception, but also the historical narrative of the company, for instance the tales a company tells about where it has been and where it is headed.[26]

At the other end of the spectrum, brand realists such as Kevin Keller claim a brand is the property of a product.[27] In this instance, the branding process is largely concerned with emergent products. That is, the image of a brand is not definitive, as an idealist would posit, rather it is "embodied in the product characteristics."[28] The brand constitutes brand image, not vice versa. For the realist, the brand comes into existence only once it finds a niche within object space, which is not to say that it is reducible to the space in which products exist, only that it survives by virtue of the presence of other products in the same set. According to this view a brand is ontologically dependent upon other brands that together constitute object space; the success of any given brand comes from how integrated and unified the various brand elements are.

Whether we argue that branding originates in the image (idealist), or the product space in which the brand exists (realist), the effect of both is what Rem Koolhaas has so poignantly described as "junkspace." We cannot grasp, remember, measure, or even code junkspace. It is chaotic and aseptic at the same time. It marks the acceleration of formlessness and mutation. As form withers we are left with a directionless, transitory, indeterminate, promiscuous, and repressive space. Ian Buchanan explains junkspace is the "residue of capitalism," a space that has been reduced to a "mass-manufactured good."[29] Junkspace no longer grants us a sense of place and, for this reason it is not only unreadable, it is impossible to navigate. Fredric Jameson declares that junkspace marks the moment when "reality begins to sag like a drug hallucination and to undergo vertiginous transmutations revealing the private worlds in which we are trapped beyond time."[30] Hal Foster observes, "as megastores govern more and more movement through cities, architecture and urbanism are more and more exposed as the mere coordination of flow."[31]

Koolhaas defines junkspace as absolutely political, insofar as our critical faculties are compromised through the experience of pleasure, comfort, and endless shopping. In fact, politics is reduced to mere entertainment in junkspace. Furthermore, and more significant for the purpose of the discussion here, the modules of junkspace are "dimensioned to carry brands," says Koolhaas, because in junkspace brands "perform the same role as black holes in the universe: essences through which meaning disappears."[32]

Brands are the floating signifiers of transcultural landscapes—mutating and modifying space without ever actually transforming it, turning the landscape into one big advertisement. So how is ecobranding impacting upon junkspace?

The new eco–Wal-Marts in McKinney, Texas, and Aurora, Colorado, are experimental stores showcasing the company's commitment to become more socially and environmentally responsible. They use new environmental technologies to reduce waste, conserve water and electricity, and minimize the energy and resources needed to maintain the store. Some examples of the new technologies include using fly-ash and slag to reduce the carbon dioxide emissions of cement; reusing cooking and motor oil to heat water; and using low-energy LED lights for signs and in freezer cases. The sustainable objective of the green store is to reduce Wal-Mart's overall environmental impact and to promote Wal-Mart as a leader in sustainable building and business practices. Charles Fishman reports that as of 2006, Wal-Mart has 3,811 stores in the United States; 1,906 of those are supercenters, with on average 16 new supercenters opening a month.[33] Three ecostores are barely a drop in the Wal-Mart ocean. That said, Wal-Mart CEO Lee Scott announced he intends for Wal-Mart to aspire to the following three goals: (1) Be supplied 100 percent by renewable energy; (2) Create zero waste; and (3) Sell products that sustain resources and the environment.[34]

Mike Duke, executive vice president and chief executive officer of Wal-Mart stores proudly explains: "As the world's largest retailer, we are excited that we can lead the way in promoting the use of sustainable building and business practices in retail and the real estate development process," going on to explain that Wal-Mart intends to share their "experiences with the industry, the general public and government agencies."[35] This is certainly a step in the right direction. However, it is important not to confuse the adoption of sustainable technologies by big-box retailers with a sustainable business model that eventually will transform junkspace. The most positive outcome of a company such as Wal-Mart supporting renewable energy sources is that it would help to significantly reduce the cost of renewable and alternative energy technologies making them more competitive in price than carbon energy options. The ripple effect being that as these cleaner technologies become cheaper and more readily available, it will be more likely that eventually communities all over the world (especially off-the-grid communities in the developing world) could be provided with a combination of renewable energies depending on their local conditions.

There are, however, a series of obvious contradictions to an eco–Wal-Mart. First, the land use pattern of the stores—located on a city's periphery—promotes urban sprawl. Sprawl places downtown business activities and the vitality of urban life under enormous stress, not to mention that it encourages commuter shopping as consumers are forced to drive to pick up the most basic household items, such as milk or batteries. One of the defining features all Wal-Marts share is their size, with vast amounts of land allocated to parking. So when stores close and relocate, it is not just the building that remains vacant but large areas of asphalt as well.

The new eco–Wal-Marts do not profess to change their underlying business model, which has resulted in hundreds of vacant stores and lots, they merely introduce the science of environmental technology into big-box design as a cost savings measure. Striving to expand as quickly as possible means many smaller-scale stores are abandoned and replaced by supercenters. It is far cheaper for the company to build a supercenter from scratch than to convert an existing store. Yet the large scale of many big-box spaces is unattractive for potential buyers and tenants.[36] The United States has an average of 31 square retail feet per person (which equates with 12.7 Manhattans) as compared to 10 square retail feet per person in the United Kingdom and 0.1 square retail feet in Africa; one of the most telling figures cited in the *Harvard Design School Guide to Shopping* is the total area allocated to retail in the United States, which amounts to 39 percent of the worldwide retail area.[37] Of this 39 percent, in 2000 U.S. Wal-Marts occupied 3.6 percent.[38] Also in 2000, Wal-Mart was the world's largest retailer with sales totaling $165 billion and its entire retail area occupying 301.5 million square feet worldwide.[39] In this context, Wal-Mart as an ecobrand is an obvious misnomer.

Any brand—whether it is an ecobrand or not is irrelevant—is premised upon the idea that the landscape (understood not just as a geographic area but a territory defined by relations formed between social, economic, political, ecological, and cultural life) is passive, meaning it is there as a medium for consumption. Koolhaas poignantly summarizes this situation using the following formula: shopping = ecology. Sze Tsung Leong reiterates:

Not only is shopping melting into everything, but everything is melting into shopping. Through successive waves of expansion—each more extensive and pervasive than the previous—shopping has methodically encroached on a widening spectrum of territories so that it is now, arguably, the defining activity of public life.[40]

It is through shopping that the market economy has come to dominate space, architecture, and life, fashioning space "in the image of the commodity" so that, as Foster points out today, "not only commodity and sign appear as one, but often so too do commodity and space."[41]

Where previously work and housing brought people together, in the contemporary world it is shopping that unites us. Junkspace is not only the effect of modernization—more specifically technology and mechanization, such as the escalator and air conditioner—it is also the effect of capitalist consumption. As Fishman observes: "Today Wal-Mart sells more groceries than any company not just in the United States but in the world."[42] In the United States, over 100 million customers shop at Wal-Mart each week and according to Fishman's calculations 7.2 billion people are estimated to visit a Wal-Mart in 2006, which translates as every person on earth (6.5 billion people) visiting once with at least half a billion visits remaining.[43] My concern in today's ecochic climate is that junkspace arises not so much from the values and actions of multinational corporate culture than as the effect of the multitude's green consumption habits.

The ecobrand uses the affective charge of sustainability culture to increase the power of the Wal-Mart brand, all the while continuing to contribute to the junkspace any activist involved with environmental and social justice issues works so hard to dismantle. The activist starts out by asking how landscapes can participate in promoting and activating ways of living that are sustainable. According to this view landscapes are dynamic changing systems. Rather than turn the landscape into a consumer franchise, sustainability culture emphasizes the creative and collaborative dimension of landscapes working across the following three poles. First, sustainability culture engages with and is informed by the cultural practices and traditions that landscapes activate and by which they are shaped. Second, it is attuned to social practices, such as informal and formal rules and regulations that promote an equitable environment, or the economic organization of landscapes into private and public land that produces and renders social organization visible and poverty invisible. Third, sustainability culture attends to the physical properties specific to any given landscape, such as natural resources, climate, orientation, infrastructure, circulation paths, and topography. In other words, sustainability culture treats the landscape as an affective process of dynamic change (what I describe later in the book as *site infrastructure*).

As Koolhaas warns, if the spatial patterning of consumption is junkspace, then for the activist the spatial patterning of sustainable design looks at how these patterns influence the dynamics of life from the biological through to the social, all the while attending to different landscape scales— micro, home, city, region, nation, and globe. One such example is when the Institute for Local Self Reliance pushed for new legislation to be introduced into Maine that would slow the expansion of large-scale retail development in the state and instead lend support to local political situations to come into being. Maine's Informed Growth Act (LD1810) applies to any store over 75,000 feet (the size of one and a half football fields). The act stipulates that an independent commission be brought in (at no cost to the town) to assess the economic impact of the proposed store on local business, employment, wages, municipal revenue, and the cost of public services; when the assessment is complete these are presented at a public hearing, at which time anyone (including the applicant) can submit evidence or testimony concerning the estimated impact of the development. The act extends preexisting Maine legislation, which gives towns the right to ascertain whether a proposed development would have an adverse impact on air and water quality and traffic, to include an assessment of the economic impact of big-box retail. The assessment is specific to the situation of each town and its community, and yet the goals are universalized around a principle of sustainable growth. The act conceives collective freedom as arising out of the singularity of each situation. Freedom, thus understood, is a local and critical activity that a community works hard to sustain.

Sustainability culture produces a radical mode of self-determination from the vantage point of the current moment.[44] And this is how artist Michael Rakowitz approaches the question of sustainable living in his work. By using the limits of design language and what it can create, he articulates the limits inherent to the state of homelessness and the laws that render the homeless invisible throughout the city. Commencing with what is immediately practical in a given situation, he uses the formal language and logic of design to generate strategies for homeless people that allow them to independently seek shelter throughout the city, affirming the unnameable Real (privatization of space, homelessness, and a homeless-free definition of the city) defining their situation. His mobile public housing project, *paraSITE*, constitutes a site-specific intervention into the privatization of public space. The homeless are equipped with a portable structure (whose shape they design) made of affordable materials (tape and plastic) that can then be attached to buildings and inflated using vented heat. The concept behind this work is that the air waste emitted from one structure is put to work to breathe life into another. As the name suggests, when the inflatable structure attaches itself to a host it feeds off of its energy and, in effect, it materially and energetically inscribes itself into the order of the city, using the city against itself so as to discover the nooks and crannies where ruptures might appear. The *paraSITEs* present and magnify seemingly unheard and unseen ways of life that already exist throughout the urban fabric.

A more recent invention of Rakowitz, *(P)LOT*, serves the dual function of shelter and camouflage, interrupting anti-tent laws in cities such as New York that prevent the homeless from setting up camp and close public space off from spontaneous forms of appropriation. He buys an off-the-shelf car cover and then builds a simple light structure that holds the cover. The wrapped frame creates a habitable space inside, but from the outside the form of the structure looks like a covered car. In this way, the *(P)LOT* transforms parking lots and metered sites into invisible urban camping grounds. Imagine these collapsible structures appropriating the hundreds of vacant Wal-Mart lots throughout the United States; moreover, what if, within the new social and environmentally responsible vision of Wal-Mart, the company architects designed vents throughout their parking lots to provide favorable temperature conditions for these makeshift shelters, actively encouraging the appropriation of empty spots at night when business slows or the store closes and the homeless need shelter the most.

Rakowitz produces protest structures that play with the power dynamic of visibility and invisibility that marginalizes the homeless. He seeks to both camouflage the homeless to provide them with a safe haven but also to render the invisible condition of homelessness visible, turning the whole notion of making the city homeless-proof on its head as the structures benignly feed off of their host's waste. Rakowitz's homeless shelters provide an interesting way of thinking about the politics of sustainability culture because they disclose the void in a situation (invisibility, homelessness, and the draconian laws producing this), and in so doing they transform our established knowledge of that situation. Similarly, sustainability culture is committed to producing a radical social and political transformation of junkspace by addressing the universal dimension of truth (sustenance, life) through the specificity of a given situation.

Another example would be the recycling and compositing work of the People Powered brand (artist Kevin Kaempf). With *Soil Starter* (2002 ongoing) Kaempf worked with local neighbors who wanted but were unable to compost their waste. Collecting and composting the waste himself, Kaempf returns the composted material back to the participants in the form of tea bags. When watered, these tea bags nourish everyday houseplants. People Powered utilizes the situation of consumer culture and branding to create a militant independence by using branding against the grain. In the hands of Kaempf, the brand interrupts junkspace as it garners local energies, channeling these toward a different economic practice; one that is creative, local, and productive as compared to an economics of profitability and corporate promotion. People Powered demonstrates that as sustainability culture is put in the service of ecobranding, it ultimately betrays this local condition, one that is open and otherwise free from the hierarchical relations multinational corporate economics advances.

In conclusion, given the amount of influence big business can have on sustainable development, it has the power to shape and change current environmental standards for business and building practices across the globe. For this reason alone, the sustainable goals of multinational corporations are certainly important. That said, there is a fundamental difference between ecobranding and the more radical practice of sustainability culture. The former implies that capitalist consumption and sustainable living somehow go hand in hand. The latter is more committed to a political conception of sustainable living, one that is not content merely to modify junkspace. By struggling over the meaning and practice of sustainability, sustainability culture seeks to transform junkspace in an effort to create an environment in which life as a whole can thrive.

In addition, it is undeniable that ecobranding is connected to greenwashing, but it is equally important to recognize how the two differ. It is not just a company image that ideologically manipulates public perception in order to nudge the brand into object space, as greenwashing presupposes; it is the energies of the social field that also invest in green consumption and sustainability culture. When this affective power is institutionalized, a series of hierarchical power relations come into being, and as I show in the second half of this book, more often than not the poor bear the brunt of the burden. Put differently, ecobranding engages with and strengthens the social power of sustainability culture, yet often it does this by managing and institutionalizing what is otherwise the open and creative state of a potentially revolutionary dynamic of affectivity producing instead what Foster lucidly observes in *Design and Crime* as a "branded subject." Ecobranding is not an instance of negative power—taking control of the individual psyche of the consumer—it is an example of affective power—empowering the social energies and affects to which sustainability culture gives rise. What is problematic is how the corporation puts these affects to work within a skewed system of power relations.

All in all, the greenwashing thesis presupposes that the greenwash misrepresents the fundamental principles of sustainability culture. Undeniably, in its focus on the coercive power of branding, the theory of greenwashing ignores the affective power of ecobranding. For this reason, it reinforces the selfsame dynamic that it sets out to undermine because as long as we understand power in coercive terms, we will remain blind to the manner in which our activism against corporate culture is an act of mutual empowerment. As the examples of the socially and environmentally responsible investor and the green-oriented consumer demonstrate, this form of economic activism may engage in practices that aspire to change how multinational corporations do business, but in the process they not only empower themselves but they empower the corporate sector as well. However, as the power base of the corporate sector is enhanced there are also no guarantees for how it will put this power to work.

The affective understanding of power helps explain how the activist social body of sustainability culture has more recently become complicit in its own repression, the reason is it mistakenly sees activism in ideological terms rather than in terms of investments of social energies. Foster explains: "Desire is not only registered in products today, it is specified there," by which he means the commodity might be mass produced but its affective power lies in how it can be personalized and tweaked to fit the profile of the individual consumer.[45] In effect, what we are buying is the feeling of power that an ecobrand image gives us, not the object itself. For this reason, as the social and material conditions of economic life take the lead, sustainability culture has not only gone mainstream, it runs the risk of being turned into a form of decoration. And where sustainability culture once was involved with constructing a new mode of mainstream activism, these selfsame political subjectivities are now on the verge of disappearing in the delirious landscape of corporate capitalism and green-oriented commodity culture. Unlike the sustainable landscapes that the socially responsible investor strives for, the ecobranding strategies of the corporate sector do not change the junkspace of capitalism; they merely modify it.

2 Green Idol

"Well does political scare ya?"

"Political doesn't scare me, radical political scares me, political-political scares me."

—Excerpt from *The Player*[1]

Yes, it's true! A tidal wave of environmental and social justice activism is hitting Tinseltown. The movie industry now trades in carbon emissions in order to produce carbon-neutral movies. The Oscars went "green" in 2007, at the same ceremony that gave the Academy Award for best documentary to a film on global warming, *An Inconvenient Truth* (2006). And movie stars are endorsing the benefits of ecofriendly living and raising money and awareness in support of the sustainability cause. In other words, the Hollywood community has gone activist, promoting a new brand of ecochic— a trendy, stylish lifestyle that is green to boot. Driving their fuel-efficient hybrids and turning up their noses at SUVs and Hummers, celebrities including Leonardo DiCaprio leverage their mass appeal to kindle the public's interest in environmental and social justice issues. But what kind of politics is being championed here? The politics of the Hollywood entertainment industry is an especially slippery area of inquiry, for if the political use of Hollywood activism is only grasped from the vantage point of ideological content then we do a real disservice to the ideological effectiveness of popular culture. This chapter examines how the political function of Hollywood activism might be understood as an affective encounter operating between the Hollywood industry (production and dissemination of movies, ideological content of films, stars, and award ceremonies) and the general public.

In respect to the Hollywood film industry and the stars who are major commodity products of that system, the common supposition is that the texts Hollywood produces are mediated by a whole new multinational and militarized economic system that seriously compromises its ability to experiment with and interrupt the selfsame system of which it is an effect. Recent empirical work has maintained that Hollywood's political function is undermined by the fact that it is "more invested in producing entertainment that is politically generic and palatable to a general audience."[2] William McIntosh, Rebecca Murray and Debra Sabia used this hypothesis to frame the parameters of their research on the connection between political characterization in Hollywood movies and mainstream political life. They studied forty-seven films according to the following criteria: films released after 1945, not made for television, nondocumentary, and included at least one main political character. Their research found that the most "striking feature" was that "most political films were ambiguous in terms of partisan politics."[3] They concluded that "most political films present commentaries on the American political system, but pointedly avoid the riskier business of commenting on partisan political issues."[4] Put differently, Hollywood texts are ideologically neutral because they strive for mass appeal.

The McIntosh study seems to sit uncomfortably alongside the findings of the *Southern California Environmental Report Card 2006*, published by the UCLA Institute of the Environment. Instead of examining the political content of Hollywood films, its focus was on the environmental politics of the industry itself. After examining the film and television industry's trade publications, such as the *Hollywood Reporter* and *Variety* magazines, it found environmental issues peaked in 1993. Then, after a brief tapering off in the mid-1990s, environmental content began to increase again in 1996 with evidence pointing to a significant acceleration from 2002 to 2004.[5]

In further support of the political content of the Hollywood movie industry, the annual awards presented by the Environmental Media Association (EMA) set out to celebrate and "honor film and television productions and individuals that increase public awareness of environmental issues and inspire personal action on these issues."[6] Some of the films that received the 2007 EMA Award for promoting green content include: *Happy Feet* (feature film), *Big Ideas for a Small Planet: "Wear"* (documentary), *Numbers: "Waste Not"* (television episodic drama), and *Bindi the Jungle Girl: "Not That Many"* (children's live action).

The EMA Awards, the UCLA *Report Card 2006* and the McIntosh study all are premised upon the idea that the politics of a text equates with its ideological content. However, feminism and postcolonial theory have taught us that representation of ideological content is never neutral and, therefore, it is always important to consider what the representation erases, as well as the deeper systems of oppression in which it participates. For instance, advocating that the movie business is inclusive, Lary May points out that Hollywood has a history of representing the values, beliefs, and norms of society in cinematic form by combining popular values, such as pluralism, with other political struggles, like the civil rights movement, in ways that have helped create avenues through which new national identities can be imagined.

Anchored in the feminist premise that identity, representation, and power are implicitly connected, May claims the representation of women and minority groups in film and television has participated in a broader cultural shift allowing for a more inclusive understanding of U.S. national identity to emerge. According to this line of argument, by representing the interests of minority groups culture has made the minority visible in the popular imagination, enabling these groups to advance away from the social margins and take a place center stage alongside everyone else. May's conclusion differs from the three examples cited above because he addresses the power structures that define how the Hollywood industry participates in a broader system of representation, prompting us to consider the politics of representation not simply as a problem of ideological content but how that content is written into the text itself. Put differently, content is not uniform. We therefore need to attune ourselves to how manifest content is symptomatic of a deeper repressed layer of content. This position alerts us to the question of whether or not the voyeuristic or narcissistic identification with the manifest content of a film can ever be political enough.

David Ingram carefully documents the appearance of environmental issues in the plotlines of Hollywood movies. He warns his readers that to define the environmental politics of movies in terms of their manifest content leaves latent content unexplored. For instance in *Dances with Wolves* (1990), cavalry officer Lieutenant John Dunbar (played by Kevin Costner) is an army deserter who befriends both a wolf and the Native Americans he encounters. The moment the Sioux chief renames Dunbar

"Dances with Wolves," his complicity in the white man's conquering of the wilderness and the Native Americans is suddenly revoked. Just as the wolf and the indigenous peoples live in harmony with the natural world, so too does Dunbar. Ingram explains how this scene exemplifies the "therapeutic tendency in Hollywood's approach to environmental concerns."[7] The traumatic history that has witnessed white man plunder and conquer nature and Native American culture is symbolically mended, as white man is reborn under a new name. Not only does the exoticization of Native Americans in *Dances with Wolves* erase their complex cultural traditions and heterogeneous history, but wild animals are also represented as "subordinate companions to human beings."[8] More explicitly, the redemptive moment associated with Dunbar's renaming disguises a deeper repression operating beneath the narrative's manifest content. Ingram explains that the film is symptomatic of a need to "renew the hegemony of the white American male, restored to innocence through mythic contact with the redemptive purity of nature."[9] Like McIntosh et. al., Ingram cautions that the Hollywood film industry's interest in promoting environmental sensibilities always needs to be moderated by a critical realism, one that recognizes the industry's "interest in promoting commodity consumption as a social good."[10]

Ingram recognizes representation is not a neutral carrier of popular values and ideas, insofar as it establishes hierarchical relations of content. What his position does not take into account is the role of the audience in producing the political function of the content. For instance, why did audiences (including Native Americans) identify so deeply with Dunbar? The film succeeded monetarily, artistically, and politically: it grossed $185 million (making it one of the highest grossing Western films in history), won seven Oscars and the Golden Globe for best motion picture, and prompted the Sioux nation to adopt Costner as an honorary member on the basis of his positive portrayal of the tribe. This widespread support points to a different layer of latent content than that offered by Ingram.[11] Adding to Ingram's thesis, I suggest that the message of reconciliation in *Dances with Wolves* is the manifest, not latent, content of the plot.[12]

Dunbar's conversion signifies a return to a more "personal and psychologically satisfying world," one in which nature, animal, and human being lived in harmony.[13] The framework for the plot constructs a time when human existence was in synch with the natural world, allowing the fragmented subject of late capitalism to enjoy a renewed sense of place and coherency. Our fascination with the life of Dunbar may have something to do with white man's redemption, but it is also a libidinally gratifying wish fulfillment. Put differently, although it is important to acknowledge that the identification with Dunbar is symbolic of white man's redemption, it is equally important to recognize this is the precondition for a deeper collective fantasy over the nature of human existence, one that is in opposition to the fragmented subject of late capitalism. Dunbar is empowered not through warfare or by virtue of how much money he earns; his power comes when he attains a place among the Sioux tribe and strikes a balance with the "natural" world. Symbolic gratification takes place through the narrative framework out of which the plot emerges (living in harmony with each other and the world around us), as opposed to the plot itself (the story of white man's reconciliation with the Native Americans).

If the political function of Hollywood environmentalism is not so much a problem of content as much as of the form of the cultural text, then we need to further examine the political function of the environmental Hollywood artifact (activist star, green Oscars, *An Inconvenient Truth*, and so forth). This can be done only after some consideration about how Hollywood has come to hold such widespread cultural influence and power across the globe. At first, one might raise an eyebrow of suspicion when speaking of Hollywood and popular culture from a sustainability point of view. The whole commodification and reification of life that the Hollywood industry participates in and actively promotes could immediately be interpreted as advancing the hegemonic system of late capitalism and militarism, of which wildlife, the environment, and the poor are all casualties. For instance, the complicity between Hollywood and the U.S. military is well known; the Air Force receives approximately a hundred film scripts a year and even promotes inclusion of its military arsenal in movies by inviting producers and directors to demonstrations of new weapons systems. Writing for the *Los Angeles Times*, journalist Jonathon Turley reports that the military has the power to censor the content of film, not to mention to stop the film from entering distribution altogether. Phil Stubb, a military liaison for the movie industry, explains: "Any film that portrays the military as negative is not realistic to us."[14] This seems to cast a shadow over McIntosh's study that argues for the apolitical character of Hollywood films. In fact, McIntosh et. al masks how Hollywood politics

operates at a formal level in what could best be described as a military service (promoting military values and industry). Political, commercial, and
military interests shape Hollywood, or as Aida Hozic describes it, "Hollyworld," which is understood to mean a production system combining the
real and imaginary in an effort to produce and disseminate "America" and
"the world."[15]

Hozic has closely studied Hollywood's rise to become a transnational
franchise, situating this growth and expansion within the context of man-
ufacturers and merchants. Initially, Universal City was a mini-city within
the larger urban context of Los Angeles, having its "own City Hall, police
department and jail, fire department, post office, school, and bank."[16] She
details how the studio was not just a place where movies were produced
but how it also exemplified social and economic modes of production. To
help make her case, Hozic quotes one of the patriarchs of the motion picture industry, Thomas Alva Edison, who pronounced in the same spirit as
Henry Ford and Frederick Winslow Taylor that "the future race depends
on quantity production."[17] The result of Hollywood's adoption of Fordist
modes of production was the rise of the studio as a controlled environment, one that could maximize production without the world outside
interrupting what was going on inside. In turn, this meant that during the
1920s Hollywood was vertically integrated with its studio headquarters
and production facilities all located in Los Angeles.

When the Fordist manufacturing model was no longer competitive, Hollywood adopted a postindustrial model of production. That is, labor was
outsourced and the reliance upon established production companies was
replaced with short-lived contractual arrangements. In this way, a more
horizontal organization took hold. For Hozic, transnational Hollywood is
inextricably bound up with the cultural logic of late capitalism. In short,
she points out how the film and television studios moved further away
from manufacturing and closer to being a mass media conglomerate operating on a global scale, one whose primary role today is branding, mer-
chandizing, financing, and the protection of intellectual property rights.

One of the difficulties of this new, unstable supply chain model is that it
is very difficult for Hollywood to institutionalize environmental practices,
and the ecological impact of this situation has recently become cause for
serious concern. The UCLA Institute for the Environment *Report Card 2006*
noted that the Hollywood film and television industry is responsible for

emitting 140,000 tons of criteria pollutants annually into California's atmosphere.[18] The report compared the filmmaking industry to six other major industries in California: petroleum, aerospace, semiconductor, apparel, and hotels. It found that "while the film and television industry in California is the smallest of the six sectors studied, it may be surprising that the GHG (green house gas) emissions are even of the same order of magnitude as in the other sectors."[19] The film and television industry relies heavily upon transport and energy consumption, such as idling vehicles, generators, machinery, and trailers that together generate large amounts of carbon dioxide.[20]

The industry has tried to mitigate this problem by employing the services of organizations such as Future Forests who calculate carbon dioxide emissions and then offset these by planting trees. For example, Future Forests calculated that the production of *The Day after Tomorrow* would generate a total of 10,000 tons of carbon dioxide. In response, the movie's director Roland Emmerich and his partners paid Future Forests $200,000 to make the production carbon neutral.[21] Similarly, Warner Brothers paid Native Energy, a similar organization, to reduce its ecological footprint for the production of *Syriana*. For the second and third films in the *Matrix* trilogy, Warner Brothers went into partnership with the Alameda County Waste Management Authority and the ReUse People to recycle 97.5 percent of the 11,090 tons of material used on the sets. Some of this material was reused to build low-income family housing in Mexico.[22]

Reducing the ecological footprint of the Hollywood movie industry also provided the motivation behind the greening of the 2007 Oscars. For the first time in the history of the Academy Awards, producer Laura Ziskin and the Natural Resources Defense Council (NRDC) worked in tandem to ensure that supplies and services for the event were chosen with the environment in mind. This included offsetting carbon emissions of the preshow; returning upholstery to the manufacturer for reuse; using paints that were low in volatile organic compounds (VOCs); making tables from wood certified by the Forest Stewardship Council; using carpeting made from 100 percent postconsumer recycled materials; recycling beverage containers; renting and reusing serviceware when possible and for the remainder using materials that were biodegradable and/or included recycled content; providing press rooms with 100-percent-recycled napkins; recycling batteries; printing programs and invitations on stock made from

30 percent postconsumer recyclable content; and even returning flowers to the caterer for reuse and/or resale.[23]

Then, of course, came the much-awaited ceremony itself. Laurie David, director of *An Inconvenient Truth*, walked to the podium to collect an award for Best Documentary. The film, starring Al Gore, the former Democratic vice president under the Clinton administration, teaches viewers about the science of global warming and offers practical suggestions for creating a future different from the dire one currently projected. After the award, to an estimated 40 million viewers Gore and DiCaprio declared:

For the first time in the history of the Oscars, environmentally intelligent practices have been thoughtfully integrated into the planning of tonight's event to make our world healthier and help combat the threat of global warming.[24]

Without a doubt the 2007 Oscars had given a fashionable face to environmental activism.

From the NRDC literature ("Shoppers Guides to Tissue Products" and pamphlets with tips on how to live green) in the Green Room where the stars socialized, which also displayed a framed letter from Ziskin outlining the Oscar ceremony's environmental program; to the signs in the bathrooms explaining the benefits of using recycled tissue; to Gore and DiCaprio's on-air announcement describing the measures taken to reduce the ecological impact of the event and directing viewers to the Oscar and NRDC Web sites for additional information; the Oscars took on a pedagogical role to reeducate the public about being more green in everyday life. As Matt Petterson, president of Global Green USA, explained: "By arriving in 'green vehicles' instead of gas-guzzling limousines, well-known individuals are helping shine the light on smart solutions to global warming: what we drive, as well as how we live."[25] Using the public's fascination with the stars, the Oscars redirected the energies and affects that make up that fascination, pointing them toward raising public consciousness of environmental issues. That is, the relationship between the Hollywood artifact and power may be grasped as a collective issue, one that is inextricably connected to how social energies, commodities, and the popular idea of sustainability become intertwined.

According to May, the Hollywood celebrity has a long history of using screen power to influence social and political life. Researching the archives of the Screen Actors Guild, May found that as far back as the 1930s "the 'stars' used their celebrity status to promote Labor Day parades and the inclusion of women and minorities in public life."[26] With this in mind, the efforts of the stars rallying around in support of environmental justice issues is no different from what May chronicles. There is something in the "connection of Hollywood to political power" and cultural authority, says May, that invites us to think about the very "real meaning of national identity."[27]

Just as the historical involvement of the Hollywood community in advancing a civil rights agenda, today celebrities are using their star power and skills in the area of popular culture to rally the public to support the causes they represent. Actor Robert Redford's scathing critique of Bush's energy policy in the *New York Times*; the television host Oprah Winfrey's foundation to support the education and empowerment of women, children, and families around the world; and Angelina Jolie's work campaigning for the International Campaign to Ban Landmines and as an ambassador for the United Nations High Commissioner for Refugees are just a few examples of members of the Hollywood community becoming enmeshed in the democratic life of public discourse. So, is there any tangible difference between stars who use their popularity to advance political causes and a democratically elected politician representing particular ideological positions?

Writing in the 1980s, when the U.S. president was former B-grade Hollywood actor Ronald Reagan, Neil Postman finds the situation of political life deplorable. In *Amusing Ourselves to Death* he argues U.S. politics became an extension of show business. Citing the amusing advice former President Richard Nixon gave to Senator Edward Kennedy on how to make a serious run for the presidency by losing twenty pounds, Postman posits that today politics is more cosmetic than ideological.[28] The content of public discourse has emptied out as it has been reduced to sheer entertainment. The more candidates of U.S. mainstream politics need to become visible throughout the mass media, the more common it is for elected politicians to resort to celebrity tactics. Sometimes this entails using the avenues of popular culture often reserved for celebrities, such as when Bill Clinton played the saxophone at the *Arsenio Hall Show* or Mike Huckabee, governor of Arkansas and a presidential candidate, appeared on Comedy Central's *Colbert Report*.

As politicians become celebrities, celebrities are becoming political and being booked by elected officials for inauguration ceremonies and recruited to attend state dinners and other White House events. Celebrities even politically endorse the ideological position of a particular politician. For instance, actor Brad Pitt endorsed Democratic candidate John Kerry in November 2004 by flying a Kerry banner outside his Hollywood mansion. Then, of course, for a string of elected U.S. politicians, their prior celebrity status or background in popular culture (in movies or sports) has provided them with the perfect apprenticeship in mainstream political life. In addition to Reagan the famed Terminator, actor, and bodybuilder Arnold Schwarzenegger was elected as governor of California; Jesse Ventura enjoyed a successful career as a professional wrestler prior to becoming the governor of Minnesota; and musician Sonny Bono (of Sonny and Cher) served as the mayor of Palm Springs.

The potency of the Hollywood community as a social and environmental activist group is not so much a result of the manifest content of the work they do as activists, namely which cause they fight for, as it is the effect of a formal relationship underpinning the fantasy of the public itself. Although culture has an indirect role to play within this process of fantasy production, the political potential of popular culture depends on how the cultural object is used. The power of the stars, for instance, lies not only in the mediated form of a collective fantasy; they also have the power to redirect social energies invested in producing a reified cultural product (namely the "Green Idol") so that that these selfsame energies of the social field can lend support to another text of populist political discourse. It is in the dialectical nature of this relationship with other political texts where Hollywood's relevance to the broader U.S. polity lies.

A case in point is DiCaprio, who after his role as Jack Dawson in the Academy Award winning blockbuster *Titanic* (1997) became a teenage heartthrob and was later described as the "new De Niro" after his role as an undercover Boston cop in the Scorsese film *The Departed* (2006). More recently, he has put his public and private persona to work and reemerged as the "Green Idol" of sustainability culture. This is much more than a mere lifestyle choice involving DiCaprio behind the wheel of a hybrid automobile, or him flying a commercial plane instead of a private jet to conserve fuel, or even how many solar panels he has on his home.[29] The example DiCaprio sets for environmentalism is one thing, but does it amount to a political experience? As a product of commodity culture DiCaprio appeals to those selfsame habits that produce the environmental crisis in the first instance—overconsumption and a weakened collective life. After all, a "Green Idol" is a commodity, one that continues to mediate the separation of the public and private seemingly without producing a different cultural discourse to the dominant one of commodification. That is, unless the "Green Idol" is appropriated not simply as the site of mediation but also as the product of mediation. This is exactly how *Vanity Fair* used the image of DiCaprio on the cover of their May 2006 Green Issue.

In 2006 when Knut, the polar bear cub born into captivity in a Berlin zoo, splashed across the front page of *Vanity Fair* alongside DiCaprio, the once-universal symbol for the world's most popular soft drink was reframed to reinforce a different message. In this new image, the cuddly polar bear of Coca Cola's "Northern Lights" commercials from the 1990s underwent a radical facelift. DiCaprio strikes a serious pose, engaging the viewer eye to eye from atop a Jökulsálón glacier in Iceland as Knut looks up at him with curiosity. The caption accompanying the image reads:

Polar bears are imperiled by the melting of the Arctic ice. The Bush administration, which has yet to decide whether to list the polar bear as a threatened species, understands the power of symbols, and has warned government scientists not to speak publicly about polar bears or climate change. Knut, the cub on our cover, was born in the Berlin Zoo. We brought him together with Leonado DiCaprio the only way we could, in a photomontage. Knut was photographed by Annie Leibovitz in Berlin. DiCaprio, no stranger to icebergs, was photographed by Leibovitz at the Jökulsálón glacier lagoon, in Iceland. Yes, we know, there are no polar bears in Iceland. If current trends continue, there won't be any in Canada either.[30]

The composite showing DiCaprio in close proximity to Knut taps into the affective power of the bear cub and the commodity value of DiCaprio. The magazine editors used the public's sympathy for Knut (symbol of a dying species born into captivity in order to survive) and DiCaprio's celebrity status as a site where radical politics could arise. In effect, the editors placed the political situation of environmental activism in the foreground at a time when the Bush administration was dragging its heels to avoid listing them under the Endangered Species Act (1973). The distinction between the latent content (the affective power of Knut and DiCaprio) and manifest content (environmental activism) of the *Vanity Fair* image therefore needs to be read alongside these historical circumstances. More specifically, these were an impending lawsuit and growing public awareness of the plight of the polar bears.

According to the Endangered Species Act, the U.S. administration is barred from any activities that may harm either the habitat or the animals listed as endangered. This means that environmentalists automatically would be able to develop a convincing case to list the polar bears under the act if they used scientific evidence stating that greenhouse gas emissions cause heat to be trapped in the atmosphere, in turn causing global temperatures to rise and the icecaps to melt. If the polar bears were listed under the act, the United States would need to lower its emissions, and we all know President Bush Junior's policy on that. Only two months into his presidency, he abandoned the country's commitment to the Kyoto Protocol, which had been signed by former U.S. President Clinton; Bush described it as "fatally flawed."[31] The Senate responded in favor of Bush, declaring that the agreement would unfairly disadvantage the U.S. economy, especially since the targets and timelines, which were binding, did not apply to nonindustrialized or developing nations.

During Bush Junior's presidency the number of species listed under the U.S. Endangered Species Act was the lowest in the history of the act, with an average annual listing of seven as compared to sixty-five during the Clinton administration, fifty-nine for Bush Senior's time in the White House, thirty-two under Reagan, thirty-eight with Carter, and fifteen for the Ford/Nixon administration.[32] On December 15, 2005, Greenpeace, the Center for Biological Diversity, and the NRDC filed a lawsuit—based on mandatory listing times enacted in 1982—to the Interior Department's Fish and Wildlife Service, claiming the U.S. administration was moving too slowly in its decision to list the polar bear as endangered.[33]

In 2006 the World Conservation Union (previously known as the International Union for the Conservation of Nature and Natural Resources—IUCN) explained the situation of the decreasing polar bear population in the following manner:

Polar bears rely almost entirely on the marine sea ice environment for their survival so that large-scale changes in their habitat will impact the population. Global climate change poses a substantial threat to the habitat of polar bears. Recent modeling of the trends for sea ice extent, thickness and timing of coverage predicts dramatic reductions in sea ice coverage over the next 50–100 years. Sea ice has declined considerably over the past half century and additional declines of roughly 10–50 percent of annual sea ice are predicted by 2100. The summer sea ice is projected to decrease by 50–100 percent during the same period. In addition the quality of the remaining will decline. This change may also have a negative effect on polar bear population size. The long-term trends clearly reveal substantial global reductions of the extent of ice coverage in the Arctic and the annual time frames when ice is present.[34]

Regardless of the polar bear's status moving from low risk in 1996 to red within just a decade, the Bush administration was unwilling to list the animal under the act because it would be an admission that global warming were a reality, a position that the oil-friendly president was reluctant to take.[35]

Then a curious thing happened. In June 2006, the North American Sustainable Use Specialist Group (SUSG) recommended the polar bear not be reclassified as "vulnerable" and used as their justification findings formulated after consultation with government managers, hunting specialists, Inuit elders, social scientists, and those living in proximity to the polar bears. Deciding to act on behalf of Inuit interests, the SUSG produced an argument that was nothing other than a distorted form of political realism, one that hoisted the flag of Inuit civil rights up the flagpole of U.S. economic interests.

Conceding that North America contains more than 60 percent of the world's polar bear population, with the majority living in Canada, the SUSG report concluded that since the 1960s, polar bear "populations have been well managed, and where depleted have been recovered under conservation and management programs that involve regulated subsistence and hunting, den site protection, bear human conflict prevention, and education programs."[36] Arguing for the importance of preserving Inuit culture, the report recognized the iconic significance of polar bears for the Inuit. With this premise in stow, the reassessment of the polar bear's status as "vulnerable" was framed as a threat to the survival of Inuit culture. Using the traditional hunter/hunted relationship, the report argues for the cultural significance of the Inuit continuing to harvest their quota of 500 bears a year. Using the premise of cultural worth, the economic importance of foreigners hunting the polar bears whose outfitter and trophy fees amount to $30,000 per bear (two-thirds of which is reported to benefit the local economy of low-income regions), the SUSG concluded the survival of the Inuit way of life relied upon the continued sport hunting of the polar bears.

Granted the United States is the country of origin for many polar bear hunters, the SUSG argued that were the animal listed as "vulnerable," the hunting rights of these citizens would cease. This would result in further revenue losses for the Inuit. In turn, the claim was made that the polar bears have survived varying climactic conditions for over 120,000 years. The conclusion was that "changes in bear population status or listing designation are response variables, not drivers of climate change."[37] On this basis, the report stated that any "new status designations of bears" would not "affect the extent or quality of sea-ice."[38]

Indeed, if the hunter/hunted relationship distinguishes Inuit culture from other cultures then the continued existence of the polar bear is obviously of critical concern to the Inuit. For this reason the Inuit set out to charge America with human rights violations for its role in emitting greenhouse gases. The press release issued in 2006 by the Inuit Circumpolar Council (ICC) group representing 150,000 Inuit made it clear that the community was outraged at not being consulted about the change in the polar bear's status. It stated: "The polar bear is central to the image of the Arctic. It's a legendary animal in our lives. It's spoken of with reverence. It's integral to our culture. To contemplate life without it is unfathomable."[39] Although concerned about maintaining their hunting rights, the press release clearly expressed distress over the impact global warming was having on the polar bear population and their habitat. Later, in 2007, the Inuit called directly on the Canadian government to recognize the impact global warming was having on their livelihood and habitat, requesting that the Inuit partner with the Canadians in an effort to take a strong leadership role in promoting more collaboration and international intervention with respect to recognizing the importance of the Arctic in global climate change.[40]

What the example of the Inuit highlights is not that they are unable to speak, which would imply that they lack agency; the point is that they are not heard. Under these circumstances, when "white people" articulate the authenticity of Inuit culture while using this selfsame representation to distinguish Inuit from non-Inuit interests, we cannot remain blind to the power relations at work here.[41] The SUSG representation of Inuit welfare seen through the prism of minority rights in effect mediates the hidden economic agenda of U.S. multinational interest and power. All in all, as the welfare of the Inuit was represented along the lines of U.S. economic interests, the real issue framing the terms of the debate (economy and power) was suppressed and presented as something entirely different (protecting a minority group from environmental activists). Accordingly, the discussion over what could be done to help mitigate the Inuit habitat from being destroyed and the polar bear population from diminishing further was no longer the focus. For this reason it is critical to shift our emphasis away from understanding what a political representation means (ideological content) to the question of how representation is used in mainstream political discourse. In my estimation, popular culture is well positioned to introduce a different text into the mix here, one that can interrupt and challenge the clarity of the dominant representation, which is exactly what the *Vanity Fair* cover image of Knut and DiCaprio did.

The composite of disparate but highly affective elements (DiCaprio and Knut) provided a different representational structure to the dominant one on offer at the time. Flirting with intangible emotions and the libidinal energies driving these the editors of the *Vanity Fair Green Issue* set out to precipitate an affective connection with the general population. The editors massaged the affects and energies circulating throughout the social field in their use of emotive imagery (a vulnerable baby bear standing on a melting ice cap) and the affective charge of "star" power (DiCaprio) all in an attempt to shift these away from the isolated subject and back into the public arena of political commitment and collective life.

Environmental politics presents a host of different representations of what environmental and social justice activism means. The example of the Inuit demonstrates the way in which interpellation can represent the subjects of environmental and social justice initiatives without agency; the resulting representation enables the dominant political agenda to subordinate the welfare and interests of a minority. In light of this situation it seems that popular culture has a crucial role to play, not so much in interpreting the content of environmental justice politics but rather in demonstrating the possibility of a political condition of representation, one that de-reifies environmental politics by putting affect and energy to work for the collective good once more. Just as Marx argues that change is the result of class conflict and fundamental contradictions operating at the level of economic life, the notion of political change is premised upon production. The trade-off between culture as production and culture as consumption has meant that even the most politically motivated movie star ultimately seems to help consolidate the very power relations their activism sets out to challenge, partially because the "star" will always be a product of commodity culture. However, this argument fails to recognize the productive value of social affects and energies, both of which provide the raw material of collective life and upon which the popularity of the star relies to maintain celebrity status. The point being, celebrity environmental and social justice activism takes those selfsame energies and affects that the masses have invested in their fascination with the stars and puts them to good use in the service of collective life.

All in all, "political" no longer scares because it camouflages and represses the ideological contradictions that give rise to what *The Player*'s Hollywood movie executive, played by Tim Robbins, describes as "political-political." Robbins's character alerts us that regardless of whether in the formal political arena, such as government, or the sphere of cultural production, which for him would be the Hollywood industry, radical politics dialectically engages with the ideological construction of its own interpretative framework (latent and manifest content), seeking to expose the residue ideological censorship leaves behind.

3   Ecovillages: An Alternative Social Organization

Founded in 1733 by British colonialist General James Oglethorpe, Savannah, Georgia, is one of America's first planned cities. Savannah is famous for its twenty-one green squares (originally twenty-four), historic architecture, Spanish moss, the well-known Savannah College of Art and Design and of course *Midnight in the Garden of Good and Evil*, the *New York Times* bestseller by John Berendt and subsequent film directed by Clint Eastwood (1997). Wandering throughout the well-organized, neatly tree-lined streets past antique stores with horse-and-carriage rides clopping by filled to the brim with wistful tourists, one could easily mistake downtown Savannah as a city with a bristling urban population and vibrant street life. However once the tourists leave, and especially when the art students clear out at the end of each quarter, the streets empty and one might wonder where all the 307,995 residents cited in the 2005 census actually live.[1] Like many U.S. metropolitan areas, the answer to that question requires hopping in the car. First, you would head south through what is commonly referred to as Midtown—a busy, dense, grungy predominantly African-American neighborhood where adhoc cook-ups take place and the smell of barbeque ribs wafts by; street vendors display a range of t-shirts and runners, and teenagers parade up and down the sidewalk or linger on porches whistling at the "bootie" strutting by. This area is slowly gentrifying as college professors and other young, trendy professionals enter, in particular the artsy Starland district that boasts a series of loft condos that have been LEED-rated (according to the Leadership in Energy and Environmental Design U.S. Green Building rating system), New York–style bakery, boutique wine outlet, galleries, and a vintage clothing store.

Continuing south on Abercorn, you would cross a major east/west artery that also functions as an invisible border between the more affluent and poorer sides of the city. Here in Ardsely Park the pace slows, the streets become greener, quieter, more subdued. Pedestrians stroll, as opposed to strutting and dancing along, and nod to one another while taking their Labradors and beagles for walks. Past the palatial brick homes and ornate iron gates, the mansions dramatically reduce in size and a series of neatly clipped front lawns, pruned hedges, and side driveways make a brief appearance. That is, until crossing another significant east/west divide: Derenne Avenue, which brings on miles of strip malls, big box stores, and car dealerships, which appear to go on indefinitely until they disappear over the horizon.

Turning right and going west your trip might stop short at Hunter Army Airfield. To proceed, you would have to park and present your driver's license and car registration and declare who you are visiting on base. If all your information checked out, you would be permitted to sign in and proceed into the compound. Heading east toward the water would eventually bring you to a different series of gates, walls, and security checkpoints—the gated communities of the American dream. Why has the downtown been evacuated? Don Luymes cites three major reasons: "the desire for social control and homogeneity, the response to fear of crime, and the maintenance of private property values."[2] All these underpin one overriding thematic: residential security.

As of 2005 Savannah had one of the highest crime rates in the country, with 146 crimes per 100,000 residents.[3] In 2006 there were 11,106 crimes reported to the Savannah Chatham Metropolitan Police.[4] For the same year, central Savannah (which includes the predominantly African-American neighborhood) had the highest crime rate, with a total of 2,607 crimes: 15 homicides, 19 rapes, 172 two robberies, 126 aggravated assaults, 573 burglaries, 1,379 larcenies, and 323 incidents of auto theft. Meanwhile, Skidaway (where the gated communities of Skidaway the Landings are situated, along with several other nongated or partially gated communities) had 1 homicide, 3 rapes, 54 robberies, 38 aggravated assaults, 162 burglaries, 533 larcenies, and 123 auto thefts, which is a grand total of 914 crimes reported overall.[5] Out of this total of 914 crimes, only 35 actually took place at Skidaway the Landings (34 larceny and 1 auto theft).[6] Many geographers, psychologists, urban planners, and developers use information of this kind to justify the establishment of residential enclaves with limited public access.

For example, using geographic information systems, geographer Kumar Naresh maps the crime patterns of Savannah in an effort to understand the spatial distribution of Savannah crime. Identifying demographic conditions and socioeconomic variables Naresh argues for the nonrandom nature of crime, concluding proximity to alcohol-serving establishments, along with locations defined by racial/ethnic segregation and poverty promote the incidence of criminal activity.[7] His findings are by no means new. Deehan, Graham and Wells, and Scribner all argue a strong correlation exists between alcohol consumption and violent behavior. Not to mention, the broader connection Naresh draws between neighborhood characteristics (which he largely correlates with African-American neighborhoods) and criminal activity that are also analyzed by Zhou, Gorman, and Horel, whom Naresh cites.[8] Naresh clearly states that his intention is to "help law enforcement agencies and local communities develop an understanding of the spatial patterns of crime and the effective allocation of resources."[9]

What sort of urban image do such crime statistics create? How balanced an impression do they provide of metropolitan life in Savannah? It is important to point out that the causal analysis typifying Naresh's work and many of the studies his own research relies upon are far from being objective or neutral. Concentrating on the connection between race, poverty, alcohol consumption, and violence, in effect works to construct poor African-Americans and urban downtown/midtown areas within a criminal epistemological framework. For instance, although Naresh acknowledges that "care needs must be taken while generalizing crime incidences to their neighborhood contexts," the irony behind the stated research goals and methodology is that the analysis constitutes a form of theoretical violence.[10] It aspires to emancipate urbanity in the Savannah metropolitan region, with the supposedly good intention of pinpointing areas of risk, all the while incarcerating some of the most historically disadvantaged members of the Savannah community: the African-Americans.

The negative representation of urban difference along racial and economic lines works to reinforce normative behaviors and a homogenous social field. This view constitutes the black body of poverty as antithetical to law and order, and it ignores the racist history behind the construction of African-Americans, one that identifies them with criminality and a way of life that has corrupted U.S. cities. Let it be remembered that out of the

$120 billion in home loans that the Federal Housing Administration (FHA) provided to Americans from 1934 to 1962, 98 percent went to white people, and the majority of new homes built with federal support also went to whites.[11] As the cities were evacuated the taxes needed to support the economy, infrastructure and environmental health of urban areas also shrank. Together these issues compounded the already historical disadvantage African-American communities encounter.

The point is, by reducing urban life to a problem of security and risk management, the violent systems such analyses set out to combat in the first instance are justified, of which the gated communities of Savannah are one effect. Put differently, one of the salient features of urban violence is the segregation and stigmatization of African-American communities as inherently aggressive and poor.[12] Theorists whose concentrated examination of rape, murder, theft, and larceny statistics in their analysis and prognosis of urban life not only construct urban life within a rationalist frame of reference, they also discursively turn crime into the definitive subject of urbanity. Consequently, they contribute to the racialized definition of the city that motivates white flight away from downtown areas. This sort of epistemological project makes urban life more available to systems of domination and management, what is more is that it produces residential planning solutions that turn their backs on the downtown, such as the gated community model does.

What exactly is a gated community? They are residential refuges, clearly bounded by a wall and often with a secured entrance (a security guard, intercom, or video surveillance) that precludes the general public from freely entering the compound without prior permission from a resident of the community. The public is therefore given restricted or, more accurately, no access to what would otherwise be public spaces, such as beaches, sidewalks, playgrounds, and parks. These are intentional communities designed to seal residents off from the messiness and perceived dangers and threats of urbanity. Typologically their distant historical roots lie in the walled cites of medieval Europe and feudal England and, like these defensive structures that were built to protect communities when under attack, the gated communities of America are defined by similar militaristic principles. Planning control and exclusivity into their layout, these settlements logically expand suburban planning initiatives that kicked in during the 1950s, which were in themselves an extension of a U.S. national defense policy aimed at decentralizing the defense industries during the late 1930s and throughout the 40s. This decentralization process saw the defense industries—bases and manufacturing plants—distributed throughout rural areas and the urban periphery. As a result, new housing developments sprung up to meet the demand of a new migration of workers employed in the defense industries, such as the defense housing developments of the Tennessee Valley Authority. The parallel lines of the defense settlement planning grid were gradually morphed into the dead-end lollipop and loop street configurations epitomizing the U.S. suburban landscape, a residential pattern that made it difficult for "nonresidents" to navigate the area. Carrying on from here, the absence of public space in suburban areas meant residential life in America was gradually privatized. However, as cars became cheaper and immigrants and African-Americans started to enter the middle class, suburbia was no longer necessarily a secure and protected domain for white America. As Edward Blakely and Mary Snyder point out, using physical space to create social place in America is nothing new, however the development type, commonly known as the gated community is "one of the more dramatic forms of residential boundaries."[13]

Citing one of the country's top national real estate developers, who in 1997 anticipated eight in ten new urban projects would be gated, Blakely and Snyder estimated at that time approximately 20,000 gated communities with more than three million units existed across America.[14] They suggested gated communities offer "a metaphor for the social processes at work in the nation's political and social landscape."[15] That is, since the 1980s the model of private ownership has increasingly prevailed at the expense of public and civic life.[16] In this view, gated communities are symptomatic of a tumultuous time in U.S. history: as crime rose, employment became unstable and labor was outsourced offshore, migration amplified, and communities were destabilized. These features of late capitalism have generated a sense of insecurity and vulnerability in the face of the future, and this has motivated many in the United States to choose a top-down approach to residential planning. Yet, as the authors are quick to point out, any "decision citizens make about the form of community settlement ultimately affects the character of that community and of the nation as a whole."[17] And popular culture has responded to this character shift in communities with sardonic irony.

The largely invisible systems of power organizing gated communities are humorously depicted in films such as Peter Weir's *The Truman Show* (1998); unbeknownst to the main character, Truman Burbank (played by Jim Carrey), his life inside the suburban township of Seahaven is one big reality show. The film cleverly employs the panopticon to structure and organize Truman's identity, as the Truman Channel producers, director, set designers, actors, and viewers all enjoy a privileged vantage point from which to view and subsequently control every aspect of his life. Although Truman senses something is amiss and at times is suspicious that someone is watching him, he cannot quite pinpoint where the problem lies. This produces a sense of restlessness in Truman to explore new horizons. In order to escape his incarcerated existence (for the majority of the film, he is ignorant of the fact that his life is a reality TV show), which the show's producer Cristof literally describes as a "cell," Truman first must overcome his insular lifestyle and his deep-rooted fear of flying and sailing.

Seahaven was filmed in and modeled on Seaside, Florida, a village designed by New Urbanism architects Andres Duany and Elizabeth Plater-Zyberk, and it epitomizes architect and critic Michael Sorkin's scathing critique of the Disneyfication of American urbanity. With only a few appearances of African-Americans (the bus driver and Truman's neighbors) and two Hispanic tourists, Seahaven is in large part racially homogenized and unmarked by the concrete material struggles of history. Filled with smilingly polite citizens, it is a theme town modeled on the principles of traditional village life. The scale is pedestrian; the overall color scheme is a bland array of pastels; all the buildings are fashioned in a neo-Victorian style and are neatly framed by white picket fences; there is no architectural variation, not even the typical U.S. brick ranch, colonial, or split-level homes. This architectural language substitutes historical depth for nostalgic effect. The model relies upon a stereotypical vision of a 1920s neighborhood as a normative planning ideal, which is then put to work as a design premise. Given the scenographic emphasis throughout the architectural vocabulary, it is unsurprising that it was later used as a Hollywood set.

Eventually, as Truman builds up the courage to cross the (artificially) stormy oceans and escape his stage-set life, he comes to occupy a different subject position backstage. Yet, the supposed reality outside of Seahaven is cast in a shadow of doubt. Moviegoers are left wondering whether in fact life backstage is just another simulation. In this regard, both subject positions (backstage and on the set of Seahaven) seem to constitute a much larger collective phenomenon, the "reality" of which is troublingly difficult to discern. In its final moments, the film unabashedly returns us all over again to the idea that reality is largely unrepresentable—one of the founding ideas of postmodernism.

The logic of postmodern space as a network of simulacra in which history comes to an end is celebrated and applied as part of the design and development vision of gated residential settlements such as those encountered on the outskirts of Savannah, and in a more cartoonish fashion in *The Truman Show*. For instance, the Landings on Skidaway is a 6,500-acre barrier island with around-the-clock security. This is a lifestyle choice the majority of citizens are barred from buying into because of financial constraints and barred from participating in because of security measures that restrict public access. The 2007 advertisement reads:

> Welcome to your own little corner of paradise—The Landings on Skidaway Island, an exclusive Savannah gated community as much known for its beauty as its personality. Located on one of the interior islands protected by barrier islands, just outside historic Savannah, Georgia, Skidaway Island retains a natural beauty undiminished by man's presence. But is it the beauty of the island that inspires the spirit of the community or does it originate from the people?[18]

In other words, escaping to the secured compound of the Landings, the distinction between the urban underworld of criminal activity and the manicured unyielding environment of the gated community is reinforced. Here traffic has slowed to a lull and homes are neatly organized in rows with their porches lining up alongside one another.

The Landings design represents stereotypes of the past. The cookie-cutout environment of suburbia has shifted slightly as the porch has replaced the side driveway and garage. Yet this has not substantially changed the conservatism endemic to suburban planning. The spatial distinction between inside (suburban and good) and outside (urban and bad) is maintained as a temporal moral order, one that is invoked between the past (good and untainted), present (confused and disorderly), and future (simply foreboding).[19] The episteme at work throughout these distinctions is an exercise of spatial power, and it is informed and shaped by the cultural dominant operating in the twenty-first century: the militarization of life.

Geographer Mike Davis in his study of the gated communities of Los
Angeles explains how exclusionary land-use policies by government, which
individual businesses and home owners replicate, are just another way of
militarizing space. He insists urban form has started to obediently adopt
what he describes as a repressive function. Examples he cites in support of
this argument are outdoor sprinklers being used to drench and deter the
homeless from sleeping in parks, restaurant owners barricading their trash
to stop the hungry from sifting through it in the hope of feeding themselves on another person's scraps, fewer public toilets, and park closures to
nonresidents (especially Asians and Latinos who tend to use these on the
weekends for family picnics) in affluent areas such as San Marino. Meanwhile, the Los Angeles Police Department, members of which now serve
on the city's design board, further contribute to the general militarization
of life of which gated communities are the effect. All these measures have
seen real estate values increase. Taken together these planning initiatives
meet a "middle-class demand for increased spatial and social insulation."[20]

As the cultural production of social life is fetishized and commodified,
the social function of a residential settlement merely becomes another feature of commodity culture. On this account, the privatized character of
the residential enclave and its stable property values are indeed testimony
to this larger system of commodity fetishism. The world of urbanity—its
racial and economic tensions, the struggle over the privatization of public
space, and zoning policies such as eminent domain that facilitate the
appropriation of urban residential areas for real estate development initiatives—is no longer an immediate reality informing the social life of gated
communities. Under these circumstances social life becomes a neutral artifact, an abstract value, a situation that masks deeper social realities and
inequalities. Basically, everyday social life in its struggle with the concrete
materialities of life and historical circumstances, have been utterly trivial-
ized in the context of a gated residential lifestyle.

We now return to the Landings advertisement. After drawing a buyer's
attention to an immutable idyllic natural landscape, the ad implies that
although close to civilization homeowners at the Landings enjoy a more
authentic lifestyle. The image consists of a composite of natural beauty,
traditional community, a recreational lifestyle, and institutionalized security. What is especially intriguing about the advertisement is the central
role "nature" plays. Where once nature was the object of scientific inquiry,

an entity to be conquered by a rational and independent thinking subject (the Cartesian cogito), during the postindustrial age of late consumer capitalism the structural significance of nature is dramatically revised so that the episteme constructing nature as Other is redistributed. As a consequence, as Buchanan explains, "nature no longer commands the same metaphysical attention it used to."[21]

To illustrate this, in German Romantic paintings, most notably the work of Caspar David Friedrich, the Enlightenment principle of human reason conquering the unpredictable forces of the natural world was overturned. Instead, as can be seen in Friedrich's *Wanderer above the Sea of Fog* (1817), the chaotic and irrational side of nature unleashes tempestuous forces, darkened skies, and a restless foggy underworld. In this picture the terrifying physical power of the natural world humbles the solitary figure standing alone on a rocky pinnacle contemplating the unpredictable charge before him. We, the viewer, take the position of this lonesome figure (whose back is turned to us) and are filled with emotion, awe, and the anxiety of our own vulnerability in the face of such turmoil. Similar to Friedrich, philosopher Immanuel Kant, who understood nature to be beyond the categories of human understanding, saw in nature the possibility for sublime experience. Later, modernists such as the Futurists, whose fascination with new technologies and the speed of industrialism, all but erased the natural world entirely from their pictures. Subjects were no longer shown in romantic natural settings; rather the body was depicted as a dynamic machine in motion. Giacomo Balla's *Dynamism of a Dog on a Leash* (1912) combines a sequence of movements superimposed over one another, the effect of which is almost an abstract representation of speed and motion. In all these cases nature was positioned as other-than-human-reason or it was the object of human domination.

As Other, nature had once been the site of possible resistance. It had the potential to confront human reason (Enlightenment), humble humanity (Romanticism), or participate in the construction of an alienated human existence (Modernism). Instead of regularizing nature by conquering or venerating it, with postmodernism the natural takes on a normalizing function. For instance, in the Landings advertisement nature enters the discursive production of the natural body and what constitutes a natural way of life. Here the category "nature" is culturally used to inscribe social life and individual bodies under a normative system of capitalist commodification.

How the six eighteen-hole champion golf courses, thirty-four tennis courts, clubhouse, more than forty miles of constructed walking and biking trails, and two deepwater marinas fit into the broader picture of "natural beauty undiminished by man's presence" is hard to understand let alone imagine.

The disjuncture between the message of the Landings advertisement and the reality of the secure physical organization of the design is telling indeed. Codifying sociality with the ideal image of a bygone natural state of affairs is the perfect example of what Jameson would describe as "nostalgia culture." That is, when culture is stripped of any referent (to the world, truth, or past) it is simultaneously condemned to the mere simulation of reality and history. As Karen Till cautions, in her discussion of New Urbanism, this constructs a moral distinction between socially good (neotraditional townships and gated communities) and bad spaces to live (subterranean criminal urbanity).[22] However, although the gated community creates an alternate reality inside its walls, it does not necessarily try to oppose the world outside. Rather, it sets out to replace it. The nostalgic effect arising out of the confluence of nature, social life, and commercial forms in the Landings advertisement emerges as historicity dies, reducing nature to a dead language. As Buchanan has so poignantly explained, this is because nature is replaced by technology, "although not technology in and of itself (just as it was never a matter of nature in and of itself), but rather what technology stands for," prompting the question of what technology stands for in the context of gated communities.[23] It stands for the "whole decentered global network of the third stage of capital itself."[24] The surveillance and security technologies common to a gated lifestyle are emblematic of this situation.

As a commercial form, technology enters the social production and conceptualization of life, transforming social life and nature into an abstract value. No longer grounded in use-value, the abstract value of social life means sociality is about the transmission and consumption of information. The more this model starts to stick, the more the ontological separation between the body and information, society and mediatization, life and consumption corrodes. Whereas Jameson might suggest it is no longer meaningful to speak of a technology/life dichotomy in the era of late capitalism, now more than ever it has become absolutely crucial that we try to do so, because what this phenomenon has produced is the increasing militarization of life.

To elucidate how gated communities contribute to the militarization of life we need to remain alert to what happens when collective life is slowly outsourced to technological systems of social organization such as surveillance cameras, security guards, other video technologies, intercoms, and so on. This issue is not just an ontological problem concerning what collective life is, or what it means in late capitalism, it is also a historical problem of how the concept of social life has been constructed through the language of defense.

The rise of suburbia after World War II came about for a variety of reasons, the most obvious being technological and economic: a dramatic rise in middle-class car ownership; FHA home loans; extensive investment in national infrastructure, such as the Interstate Highway System, made commuting easier; air conditioning, which reduced the heat island effect of vast expanses of asphalt; and, where previously urban areas provided a rich source of entertainment, when television entered the average U.S. home, families were entertained from the convenience of their own living rooms.

Meanwhile, during the 1950s a militarized undercurrent also shaped the residential development patterns. Let it not be forgotten, as mentioned earlier, that the suburban development model has its roots in U.S. defense settlement planning initiatives. For example, when Albert Kahn built the Ford Motor Company Willow Run bomber plant to manufacture B-24 bombers approximately thirty miles out of Detroit, the Federal Public Housing Administration responded with a call to build a new defense settlement—Bomber City—to house the workers.[25] Although the plans were blocked by Henry Ford, who feared the influx of workers into Washtenaw County would bring union sympathizers to the area and disproportionately change the political landscape, the master plan for a series of single-family homes uniformly organized went on to influence the design work of Eliel and Eero Saarinen's work in Detroit.[26] To accurately gauge the significance of this conflation of militaristic principles in the everyday life of America around this time, it is worthwhile quoting Senator Joseph R. McCarthy's Wheeling Speech to President Truman, "Speech on Communists in the State Department" (February, 1950) at length:

Five years after a world war has been won, men's hearts should anticipate a long peace, and men's minds should be free from the heavy weight that comes with war. But this is not such a period—for this is not a period of peace. This is a time of the Cold War. This is a time when all the world is split into two vast, increasingly hostile armed camps—a time of a great armaments race. Today we can almost physically hear the mutterings and rumblings of an invigorated god of war. You can see it, feel it, and hear it all the way from the hills of Indochina, from the shores of Formosa right over into the very heart of Europe itself . . .

Today we are engaged in a final, all-out battle between communistic atheism and Christianity. The modern champions of communism have selected this as the time. And, ladies and gentlemen, the chips are down—they are truly down . . .

As one of our outstanding historical figures once said, "When a great democracy is destroyed, it will not be because of enemies from without but rather because of enemies from within." The truth of this statement is becoming terrifyingly clear as we see this country each day losing on every front.[27]

During the early stages of the Cold War, citizens regularly participated in drills designed to better "prepare" them for nuclear war. This occurred regularly in both public and educational settings. The arms race between the Soviets and the United States was in full swing during the 1950s; and the Red Scare made it dangerous even to share intellectual sympathies with communism. It was not just new technologies and available capital that prompted U.S. white populations to leave urban areas, it was also the imminent threat of the Atom Bomb, which for maximum effect seemed more likely to be used in densely populated urban areas. The pattern of white flight into the suburbs and urban decentralization was motivated by these deeper anxieties of impending nuclear war.

As Reagan came to power a new image of a fortified America taking security into its own hands with adroit confidence prevailed. Interestingly, at this time gated communities started to replace the suburban residential pattern. Social exclusion took on an entirely different dimension as new surveillance technologies were used to shift the definition of social life; this helped create a different kind of social enclave to that of suburbia. On the one hand these were, and still are, isolated residential islands. Nonetheless, on the other hand the gated community continued to economically participate in and even reinforce the logic of late capitalism. In addition there is a militarized face behind this paradoxical organization of space—separate and at the same time fully implicated in the dominant economic system and burgeoning culture of militarism.

After having suffered the psychological blows of the Vietnam War, the United States in the 1980s, under the presidential administration of Reagan, was becoming increasingly patriotic. Military life figured prominently in popular culture, with *Top Gun*, starring Tom Cruise, coming in as the number one picture in 1986, its box office takings exceeding $176 million.[28] Andrew Bacevich argues that *Top Gun* sanitized warfare; the modern warrior lifestyle became tangibly hip: "Warm California sunshine, hot motorcycles and classic cars, leather jackets festooned with military patches and worn as fashion accessories, sleek-bodied aircraft flown by sleek-bodied men, a plentitude of beautiful women . . . "[29] Similarly, Bacevich notes that at around the same time military techno-thrillers, such as Tom Clancy's 1984 *The Hunt for Red October*, became bestsellers. During the 80s the widespread attraction for military life in popular culture and the moral high ground that Reagan restored to American's men and women in uniform provided the backdrop for gated living patterns. As U.S. military power was popularized under Reagan, this generated an enthusiasm for a militarized lifestyle throughout the social sphere, one that motivated the design and development of highly secure residential complexes, all of which shared the typological features of the military compound. Essentially these include: an isolated location, a road leading to the compound that functions as both a bridge to the outside world and a line of demarcation, and a tightly knit uniform settlement pattern of single-family homes secured together by a definitive boundary or fence.

Given this history, it is unsurprising that ecovillages in America had started to come to the fore during the late 1960s and 70s, when military life was culturally regarded as morally bankrupt and the Vietnam War was certainly becoming more and more socially unpopular. For example, in 1971 *The Farm* in Tennessee was founded, declaring that its spiritual and ideological roots lay in the hippie movement that had cropped up largely in reaction to the war; or *Twin Oaks* in Virginia, founded in 1967, whose main aim was to "produce better socialized human beings through patient education of the children"; or even the anthroposophical underpinnings of the community at *Kimberton Hills Camphill* in Pennsylvania, founded in 1972.[30] Although the ecovillage seems to share a similar principle of exclusion from the mainstream with the gated community, where it differs is in its approach to social life. The ecovillage is the very antithesis of the self-defensive social organization defining the gated community. While by no means perfect, it is an experimental and creative response to the question of how sociality works, as well as a direct challenge to the militaristic principles underpinning the organization of social life.

Taking an ecological approach to the practice of social and individual communication, dealing with diversity and the conflicts that difference may produce is understood as an integral part of a peaceful social life and vibrant community. As David Orr notes: "In a well-designed community, people would know quickly what's happening, and if they don't like it, they know who can be held accountable and can change it."[31] Although an example from without the United States, *Zegg* in Germany was founded on this premise, using the idea that a community needs to start out with conflict in order to understand how to be a community. Instead of trying to lock conflict out of the organization of the community, in the way that a gated community does, it embraces pedagogical and social effects. The ecological principle of conflict resolution claims: "People actually become closer when they work through their conflicts."[32] What this means is that unlike the gated community, which is crisis-driven—locking out potential sources of social conflict, struggle, and difference—*Zegg* is strategy-driven—structure specific, content specific, and conflict specific.

An ecovillage can be a farm—*Sunrise Farm* in Victoria, Australia (founded in 1978); a suburb—*Ecotop* in the outskirts of Düsseldorf, Germany (1995); a newage commune—*Findhorn* in Scotland (1962); a social forum based upon the premise of conflict resolution—*Zegg* in northern Germany (1991); a religious community—*Old Bassaisa* in Egypt (c.10,00 years ago); a Kibbutz—*Clil Ecovillage*, Israel (c. 1973); or even a squatter settlement—*Wilhelmina* in Holland (early 1990s). Generally speaking, it is a semi-self-sufficient, human-scale, cooperative, sustainable settlement that integrates all the primary facets of life—sociality, alternative economics, food production, energy, shelter, recreation, and manufacturing—with a sensitivity toward the environment and its natural cycles. It is important to point out that the ecovillage is not simply another residential settlement model, because its social fabric does not always consist entirely of permanent residents. There are short- and long-term residents as well as visitors who just pass through. In this way, its social organization is far more malleable than what is commonly found in other intentional communities, such as gated environments or even many other cooperative housing schemes.

Starting out with the premise that economics is creative, the ecovillage does not set out to eradicate money, although it does try to reinvent how money is used. Jan Martin Bang clarifies that in the "real economy money is a tool which can create. It is the measure of energy which circulates within the economy. Money is the unit, economy is the process. Economics consists of distribution, products, values, relationships."[33] Intent on revitalizing local economies as a way to generate economic diversification and enhance self-sufficiency, many ecovillages develop new relationships between consumers and producers. More often than not this means a whole new approach to farming, such as practices anchored in biodynamic farming techniques and principles aspiring to "increase the fertility of the land and enhance the soil."[34] These work with natural cycles like seasonal changes in climate.

Recognizing the importance of organic farming, the principle of benefiting the local economy and encouraging sustainable land-use that underscores ecovillage agricultural practices can be distinguished from more mass-produced methods of organic farming.[35] In 2006 Wal-Mart, America's largest grocery retailer, proudly declared it would offer consumers more organic products for less money. The food will be farmed without using pesticides, antibiotics, or chemical fertilizers. Though the effort to become more "green," as outlined in chapter 1, is part of a growing trend among the corporate sector, how much this really benefits the producer is uncertain.

In his practical guide to the design and development of ecovillages, Bang explains:

Agriculture today is about power, not feeding people. It is an industry . . . which takes responsibility from people and makes them dependent and powerless, turning that power into profits for the few, essentially multinational corporations.[36]

Needless to say, Bang is highly suspicious of the revolutionary power of organic farming. The industrialization of organics that companies such as Wal-Mart are now involved with is philosophically antithetical to regenerative principles at work in a more expanded definition of economics. The agricultural practices occurring in ecovillages set out to empower people and communities to feed themselves. It follows that such a focus is implicitly pedagogical. In Orr's view this means "[G]rowing food on local farms and gardens, for example, becomes a source of nourishment for the body and instruction in soils, plants, animals, and cycles of growth and decay."[37] In this context farming is understood as having a socio-pedagogical function, and agricultural labor is believed to be an inherently creative pursuit. This is not just a sustainable approach to land-use, it entails an entirely new conception of social life, tapping into prepersonal affects, energies, and forces—climate, the will to work, natural cycles, sustainable waste treatment, local economic relations, and so on.

The social life of an ecovillage is open to a multiplicity of meanings and social conditions and is the effect of a flexible organizing principle. In comparison, the meaning of social life the gated community model advances is one of paranoia, defined through a rigid and isolationist urban pattern in response to the perception that the outside world constitutes a threat. Although gated communities appear to break away from the forces and power relations defining contemporary urban life, they reinforce the inequalities such power relations produce, along with assimilating and strengthening the economic and physical realities of the military-industrial machine. In contrast, with the ecological premise underpinning the ecovillage way of life, capital functions in an activist way because it is not wholly defined by capitalist modes of production.

If gated communities attempt to resolve the struggles indicative of diverse economic and racial groups living together in an urban context by privatizing and militarizing social life, the ecovillage endeavors to embrace these contradictions as a way of life and source of conflict resolution. Although certainly a more progressive model than that of the gated community, because it tries to resolve such tensions collectively, the ecovillage still produces a largely conservative response to the urban realities of social and economic contradiction, in that it too relies upon a model of private ownership and the negation of urban life. However, in defiance against the militarization of life, and by attempting to address the problem of social conflict within its very organization and practice, the ecovillage is much more than a reactionary response to the increasing militarization of U.S. life because it offers an alternative. It advances another way of life, one that is premised upon conflict resolution, social consensus achieved through inclusive discourse, and a profound respect for the environment and local economy. To summarize, the gated community produces social rigidity. Whereas the ecovillage aspires for social resilience, and herein lies the sustainable focus of the latter.

4 The Greening and De-Greening of the White House

I never forget that I live in a house owned by all the American people and that I have been given their trust.
—President Franklin D. Roosevelt[1]

Concerned with how to live a healthy, responsible, and ethical life, the practices and theories sustainability culture propounds pose a whole gamut of questions that address our relationships to each other, ourselves, the world, and the future. As this system of knowledge gains influence it also raises a series of anxieties about the future and the tally of environmentally reckless behavior to date. Some of the questions this prompts include: What risks are involved if we do not act immediately to reverse or even slow the effects of global climate change? What is the right path to follow so that future generations do not live in a wasteland we are in part responsible for creating? How should we live today so that our children's children can continue to thrive? What does sustainable living entail, and does it mean the same thing for everyone? These are all relevant and timely questions to be asking, but they largely invite a similar response— whether or not the environment and future generations have an intrinsic or extrinsic value. What these questions do not invoke discussion of is the manner in which sustainable design and theory is a system of knowledge that shapes subjectivities. Essentially, it is a discourse that marries science, technology, aesthetics, and philosophy in a system of power relations. This chapter charts the way in which sustainable design and construction in the United States participates in that discourse and, in turn, how the U.S. administrations have appropriated sustainability culture over the years in an effort to advance national interests abroad. The question framing this discussion is: How have U.S. presidents used the greening and de-greening of the White House as part of a broader exercise aimed at legitimating federal power and military policy?

To begin, the design of the White House by Irishman James Hoban began in 1792 but was not completed until 1800. Although President George Washington (1789–1797) chose the site at 1600 Pennsylvania Avenue and laid the first cornerstone of the building in October 1792, the first president to occupy it was John Adams (1797–1801). On August 24, 1814, the White House was burned to a shell by the British during the War of 1812 and then rebuilt from 1815 to 1817. More than a hundred years later, another fire struck the building, this time only in the West Wing. Designed by Alfred Mullett in the French Second Empire Style, work on the Old Executive Office Building (OEOB) was completed in 1888.[2] Initially designed to house the State, War, and Navy departments, today the OEOB has more than 550 offices extending over approximately 600,000 square feet. In 1902 both the East and West Wings were added. Initially, the East Wing was intended to be a museum, but during the expansion in 1941 World War II broke out, and the area was used for other purposes. Today, visitors enter the White House through the East Wing prior to arriving at the ground floor of the Executive Residence. The West Wing, initially designed to be a temporary building, was expanded seven years later to become a more permanent structure through a major renovation in 1934. Later during President Harry S. Truman's term in office (1945–1953), and as a result of structural weaknesses, the Executive Residence (all except the third floor) was gutted right down to its original sandstone walls. From 1948 to 1952, it was completely rebuilt and all in all the main sections of the White House complex now include the Executive Residence, the East Wing, the West Wing, and the OEOB.

The Executive Residence was once a model of sustainable architecture with natural ventilation and a passive heating and cooling system. The East-West orientation allowed for passive solar heating and the effective use of daylight. With high ceilings and large windows and situated on the edge of a hill, the rooms allowed in breezes that would continue on to blow through the hallways, naturally cooling interior spaces. The granite structure of the OEOB contained air channels and slots under the windows to allow fresh air to enter and circulate throughout the building. The open stairwells in the corners of the building had a stained glass dome at the top with vents that could exhaust hot air out of the building as it rose. Furthermore, throughout the complex skylights allowed for natural daylighting. Finally, the First Ladies' garden provided organic produce for the Executive Residence, and at one time the grounds had been used for grazing livestock. As Ted Shelton points out, over time the White House absorbed a whole host of technologies, such as running water in the 1830s, gas lighting in 1848, electric lighting in 1891, and the installation of office computers beginning in 1978.[3] Additionally, over time many of the building's sustainable features were either compromised or removed completely. Windows were painted shut, skylights covered. And from 1888 on the passive cooling system of the OEOB was slowly unraveled to the point that when President Bill Clinton took office in 1993 it had become completely inoperative. When Clinton announced his Greening of the White House initiative (also in 1993), the cooling system consisted of 782 window air conditioning units, 100 package units, and 3 chillers.[4] Shelton says that "adoption of a technology by the White House" often signals when "technology has moved beyond the realm of the merely experimental" and has entered common usage.[5]

The White House has many functions: office, residence, museum, recreational areas, service and maintenance areas, special events, ceremonial occasions, and symbolic. As an ornate version of Palladian architecture, the symmetrical proportions of the White House, whose elevation resembles a Roman temple façade, is a potent symbol of the presidency, democracy (U.S. style), historical continuity, endurance, and national identity. This symbol articulates an image of national lineage extending to the founders of U.S. independence (Washington, Jefferson, Adams, Madison, and so on) and democracy (The Declaration of Independence, the U.S. Constitution, and the Bill of Rights). For these reasons, the fundamental premise for architects involved in rebuilding, conserving, or renovating the historic architecture and grounds has been to maintain the continuity and coherency of this image. If the image of the White House represents the heart of national identity and the historical origins of U.S. democracy, what does this process of framing leave out of the picture?[6]

Anyone taking a walk through the streets of Washington, D.C., cannot help but notice the deep social and economic inequalities characteristic of the surrounding landscape of the White House complex. The destitution of the homeless lingering in parks waiting for soup kitchens to open, along with the poverty of the surrounding slum districts, may be an unpopular and uncomfortable reality for most, but others might argue this vision of America is certainly a more honest snapshot of everyday life than the perfectly kept White House grounds. The messy and insecure realities endemic to poverty are a stark contrast to the orderly image of the White House on the oval emblem behind the press secretary's lectern or the 29-cent stamp from 1992 that shows it in detail alongside the U.S. flag.

If democracy, national unity, and identity are integral to the symbolic content of the White House, then the subsequent inscription of an architectural grammar through which to articulate this symbolic content is the moment when politics makes an appearance. Put differently, how the White House is bound to content is an important question that needs clarification, because as it is bound power is exercised. Although we may perceive the building and surrounding complex to have an origin in the founding of U.S. democracy, this image has a history.

For instance, under Jimmy Carter's administration (1977–1981) new environmental technologies not only scientifically produced knowledge of national independence and economic strength, but when sited on the White House, the categories used to establish this knowledge—a courageous, autonomous, and democratic nation—became the conditions for a new form of U.S. militarism. For Carter the economic impact of rising oil prices and America's increasing dependence on Middle Eastern oil were the main reasons U.S. confidence had taken a beating. In his energy speech on April 18, 1977, he declared his administration would teach a new, more energy efficient way of life by example. In support of this goal, he promised to establish a comprehensive, predictable energy policy; institute environmental protection and conservation initiatives; and roll back the crippling embargoes he saw as the main threat to national security. In turn, he asked citizens to implement measures to save energy, all the while advising they consume more plentiful energy sources in place of scarce fuels, which he explained needed to be conserved. This, he believed, would strengthen the economy overall.

Carter was careful to ask every individual and national organization to equally accept the burden of what he saw as a realistic and necessary series of sacrifices. Above all, in the spirit of true liberal democracy, he clearly stated no one would enjoy an unfair advantage over anyone else, declaring "citizens who insist on driving large, unnecessary powerful cars must expect to pay more for that luxury."[7] In his "Crisis of Confidence" speech in June 1977, he connected his warning of a national energy crisis to an overall loss of confidence plaguing the U.S. economy, social fabric, and cultural life. In this way, he turned his attentions to one of the most enduring symbols of democracy and national identity: the White House. Acting upon his stated mission to have 20 percent of U.S. energy coming from solar power by the year 2000, a series of solar heating panels were installed on the roof of the West Wing. As it was transformed into a more energy efficient body, the building simultaneously underwent a radical process of reinscription.

Carter's new solarized image of the White House set out to restore a sense of presidential leadership and confidence in the country's future. This was a time in U.S. history when the average person felt an insurmountable divide between citizen and government, as films such as *Dirty Harry* (1971) starring Clint Eastwood clearly depict. Eastwood plays a quasi-vigilante cop who just wants to get the job done. He continually must deal with an incompetent police force throughout the film, highlighting a societal loss of faith in law enforcement, exemplifying the general mood of pessimism, and indicating a sense of public distrust that the government and government officials can actually be effective. Carter was determined to stitch the people and government back together and reinvent America on the back of 1950s' family values. It is worth noting that Carter was a self-described born-again Christian from the Bible Belt of the Deep South. He took office with the promise that he would ensure any decisions his government made would be "designed to honor and support and strengthen the American family."[8]

In adding solar panels to the White House, Carter produced an image that could absorb and neutralize deeply ingrained social anxieties that had started to circulate as a result of the United States's poor economic performance. Using the notion of sacrificial labor, the image of a solarized White House simultaneously inscribed this particularly Christian version of national work within democratic terms. That is, each and every individual and institution would be involved equally in the sacrifices needed to revitalize the nation. The labor of the average citizen, the corporate executive, and the President all used the same criterion—thrift—in their work to save the country from the impending gloom of an energy crisis. Hence, as renewable energy was bound to the White House it not only produced a discourse of power, it also inaugurated a new mode of labor.

With the addition of the solar panels, the White House came to embody a newfound sense of sovereignty and autonomy. Carter explains:

This intolerable dependence on foreign oil threatens our economic independence and the very security of our nation. The energy crisis is real. It is worldwide. It is a clear and present danger to our nation. These are facts and we simply must face them.[9]

Although everyone, regardless of social status, had to sacrifice, the energy that went into such efforts also found investment in another ethos: cultural militarism. Carter himself proclaimed:

Energy will be the immediate test of our ability to unite this nation, and it can also be the standard around which we rally. On the battlefield of energy we can win for our nation a new confidence, and we can seize control again of our common destiny.[10]

The solarized White House was American cultural militarism in disguise.

In the opening to his April 1977 address to the nation Carter directly bound renewable energy to an act of war, stating, "With the exception of preventing war, this is the greatest challenge our country will face during our lifetime." Further on in the same speech, he proclaimed his energy plan was the "moral equivalent of war."[11] As he saw it, the only way forward was to rebuild national unity and confidence by winning the "war on the energy problem."[12] The overall conclusion was that it was not political and civil liberties that were the greatest threat to U.S. democracy but a loss of confidence in the future, which in turn was the result of being too dependent upon the Middle East for oil.[13]

Unlike the conclusion Bacevich reaches, Carter's preference for conservation and renewable energy, over and above that of overt military force, may not be testimony to him being one of the "least militaristic of recent presidents."[14] Carter's focus on energy can also be viewed as part of a much larger strategy designed to stop the Middle East from carrying too much influence in U.S. economic and spiritual life, of which the outcome was a newfound social investment in the importance of U.S. military might. Bacevich, however, is right on the mark when he explains that Carter, in fact, "created the conditions for the militarization of U.S. policy that was to come."[15] Without an energy problem, the United States would no longer be incarcerated by its dependence upon foreign oil—rather the Middle East would come to constitute an absolute space of exception—an "extreme form of relation by which something is included solely through its exclusion," and it is worth noting that over time this was to become the primary object of U.S. militarism.[16]

The moral force of Carter's environmental command to engage in sacrificial labor, consists in the command preserving itself in relation to what it excludes: sovereignty concomitantly ruled over the Middle East as the Otherness of the Middle East was interiorized. As the president himself said, the goal was to create an "energy secure nation" through sacrifice, which was none other than a form of militarized national labor.[17] The solarized image of the White House was neither political on the basis of how it addressed a given sociopolitical issue; nor was its politics the result of how it represented a particular ideological position or social struggle. It was political insofar as it refused to function in either of these ways.

Flanking the roof of the White House with solar panels produced a series of political limits: American/Middle East, democracy/lack of national confidence and purpose, sacrifice/wastefulness, renewable energy/national vulnerability. These conditioned how labor in the context of society and politics was perceived, demarcating a specific role and function to the image of the White House, which together were implicated in producing a common definition of U.S. democracy. However, in designating a courageous national subject willing to make personal sacrifices for the benefit of the collective, an outside to the White House image was also created. The labor of the oil-rich Middle East was its unseen underbelly. Although the solar image of the White House did not take the place of a much larger political struggle going on at the time—rising inflation, economic stagnation, high interest rates, rising cost of oil, and unemployment—it certainly influenced and even informed that antagonism.

When Carter was defeated and Reagan entered office (1981–1989), the struggle over the image of the White House continued. Under Reagan, the country's future was not one of limited opportunities, nor did its economic and military might have anything to do with energy conservation. Conservatism had taken the place of liberalism, and the focus shifted away from governmental involvement and regulation of the market to individual responsibility, free markets, and low taxes. Reagan optimistically took the stage also promising to reintroduce 1950s' family values into U.S. life. For some commentators such as James Davison Hunter, who in his groundbreaking sociological study of the country's cultural wars at that time, explained that in the 1980s the United States was defined by a power struggle taking place between the religious conservatives and the secular humanists.[18] For Hunter this situation resulted in the realignment of U.S. culture and politics.

Conversely, in the 1960s U.S. society and culture saw increasing liberalization. It was the decade that bore witness to the sexual revolution; the expansion of individual rights for gays, blacks, and women; and culminating in January 1973 with the U.S. Supreme Court's legalization of abortion in *Roe vs Wade*. During the 1970s a shift toward the Right began, such that 46 percent of people in the United States identified themselves as conservative in 1963, compared to 51 percent in 1969. Similarly, during that same time period, the percentage who considered themselves liberal fell from 46 to 33.[19] By the 1980s the ideological and moral separation along orthodox and progressive lines sharpened, further polarizing U.S. society. Hence, when in the summer of 1986 Reagan removed Carter's fully functioning solar panels, it was not simply because the "panels didn't function as well as hoped."[20] In fact, the panels went on to provide hot water for the cafeteria at Unity College in Maine for another twelve years. But the majority of people did not want to be reminded of the anxiety they had felt during the Carter administration.

For Shelton, the removal of Carter's solar panels gestured to a new optimism pervading U.S. life. After years of rising oil prices beginning with the Arab oil embargo in 1973, the situation had started to change. By 1986 oil prices had dropped considerably, restoring a sense of confidence and quickly supplanting the doom-and-gloom scenario Carter had described in 1977. Shelton quite rightly notes that popular reality had shifted because the population could also afford to redirect its attentions away from solar power; the implication of this argument being that subjectivity is the effect of social forces and energies finding a different investment, not a result of a specific ideological position.[21] During the Reagan presidency it was consumption, not sacrifice, that came to be the dominant mode of labor defining U.S. society and its economy. As Reagan himself said in 1987: "It's true hard work never killed anybody. But I figure, why take the chance?"[22] America was now seen to have an abundance of resources to the farcical point where trees were perceived as polluting more than factories. Free market capitalism came in full swing and tax cuts abounded. Culturally the situation translated into a form of egocentricism as designer labels were in, the art market boomed, and the new mantra "shop till you drop" took on mythic proportions—defense spending during Reagan's two terms in office came in at $2.7 trillion.[23]

The public's perception of an ineffectual military set the tone for a policy of mammoth military spending during the Reagan years. Bacevich provides a compelling argument that Carter gave the impression he took the men and women in uniform for granted, never clearly acknowledging that they were also a source of national unity and strength.[24] Further, the subsequent embarrassment of military equipment failures during the covert rescue operation of the U.S. hostages in Tehran that Carter had ordered on April 24, 1980, had, in the words of Bacevich, "persuaded Americans that the enfeebled state of the armed services had become intolerable."[25] Carter had appealed to the average citizen to make sacrifices in a collective effort to recover a sense of national confidence and unity, but Reagan emphatically rejected this idea. Carter had drawn a direct correlation between U.S. dependence on Middle Eastern oil and the debilitating lack of U.S. confidence, but Reagan associated the same lack of confidence with a weakened military.

Reagan used the image of the soldier, instead of solar energy, as a symbol of national pride and patriotism. During the presidential presentation of the distinguished Medal of Honor awarded to Master Sergeant Roy P. Benavidez on February 24, 1981, Reagan used the occasion to publicly represent the Vietnam War as a noble cause, replacing the dominant narrative of the time that had seen it as a shameful stain on U.S. history. Bacevich notes that the president established "support *for* 'the troops'—as opposed to actual service *with* them—as the new standard of civic responsibility."[26] Additionally, Reagan assured the average citizen that under his watch, reviving U.S. military might and its position in the global community would not require reintroducing the draft (of the Vietnam War years); instead of involuntary sacrifices, all that was needed was a change in attitude about how the military was perceived.

On Earth Day in 1993 President Clinton (1993–2001) announced his Greening of the White House initiative. Capitalizing on the idea of the White House as the people's house, he confidently declared that before he could ask citizens to make the best possible environmental choices in their own homes he, as president, first must lead the way in his own home. His proposal was to use the White House complex as a model of efficiency and waste reduction. The speech had an obvious precedent in Carter's February 2, 1977, nationally televised fireside chat; in his sweater

cardigan, Carter had appealed to the conscience of every citizen to conserve energy by lowering their thermostats to 65 degrees Fahrenheit
during the day and 55 at night. Both used the White House to give visibility to the issue of sustainability. Whereas Carter had called upon the
nation to make sacrifices, Clinton saw his environmental initiatives as a
reason to celebrate: "We are challenged here today not so much to sacrifice as to celebrate and create. I've challenged Americans who are young
in years or young in spirit to offer their time and their talent to serve their
communities and their country."[27]

Clinton redirected the power already invested in the sustainable image
of the White House by both Carter and Reagan; however, he did this
largely in contrast to Carter's solarization initiative. As Shelton's study
advises, "it is important to note, in contrast to Carter's rhetoric, that Clinton's vision encompasses not only energy but also explicitly environmental concerns."[28] As noted, the White House complex once had included
several sustainable features, all of which the Clinton Greening initiative
set out to reinstate and enhance. In an effort to maximize the thermal
integrity of the building shell, the roof of the White House was replaced.
All but 5 percent of incandescent light bulbs were replaced with compact
fluorescents, saving 1,600 kilowatt-hours per year.[29] Some of the other
achievements include: a comprehensive recycling program, rehabilitation
of the historic skylights, replacement of 90 percent of the windows with
energy efficient double-paned units, replacement of approximately 80 percent of the old air conditioning units with more efficient models that also
had timers to automatically turn them off between 10 p.m. and 5 a.m, a
chlorofluorocarbon (CFC) management system, replacement of chillers
with high-efficiency units that did not use CFCs, asbestos removal and
installation of new insulation on pipes, use of paint low in VOCs in 60
percent of the rooms, planting of native plants in place of nonnative species, composting of trimmings, and use of organic fertilizers.

Most important, these improvements reduced annual atmospheric emissions by approximately 845 metric tons of carbon.[30] In March 1996 the
estimated savings for expenses associated with energy, water, solid waste,
and landscaping for the White House was more than $150,000 annually. A
year later these savings increased by another $150,000, making the total
savings $300,000 per year. As discussed below, in addition to being a cost
savings measure, Clinton's Greening of the White House influenced the

behaviors and habits of the White House's staff and maintenance workers as well as the collective body of geopolitical power.

Influenced largely by the perspective of his Vice President Gore (who had held the first congressional hearings during the 70s on global climate change), the future in Clinton's Greening of the White House initiative was seen to carry the power to change human and nonhuman life. As the building was greened, so too were the habits of its staff, who were instructed to turn off their computers when they left work in order to save energy and promote security. A no-smoking policy was enforced in an effort to improve indoor air quality. A staff-education project began in 1993 with the aim of reducing paper consumption, conserving office supplies, and limiting the use of disposable items. Iniatives ranged from copying monthly accounting reports on both sides of the paper to a transit subsidy program—Metrocheck—encouraging eligible employees to use public transportation instead of automobiles.

The call to protect the environment and pave the way for long-term sustainable forms of governance was largely shaped by the notion of U.S. exceptionalism, an idea that had prevailed since the war against Nazism in Europe and later during the Cold War. When the Berlin wall came down in 1989 and the Soviet Union subsequently collapsed, a new geopolitical identity was inaugurated. As the U.S./Soviet struggle ended, the United States occupied an unrivaled position as the world's superpower. The triumphal title Clinton inherited as he came to office may have been "the leader of the free world," but under the Clinton administration the U.S. armed forces worked harder than they ever had during the Cold War (except during the Vietnam and Korean wars). President George H. W. Bush had deployed the military fourteen times abroad. Reagan had used them internationally on seventeen occasions. Meanwhile, Clinton's tally from January 1993 to December 1995 came in at twenty-five international deployments.[31]

Clinton's commitment to long-term initiatives to safeguard the environment were not limited to the protection of the national interest, they also became part of a larger crusade to save the world from impending crisis. The principle was clearly reiterated in his speech on April 26, 1993, when he announced: "as long as I live and work in the White House, I want Americans to see it not only as a symbol of clean government but also [of] a clean environment . . . "[32] He went on to clearly indicate the importance of environmental stewardship and national security, saying that the policy of environmental protection "is a legacy of America's efforts to defend our people and the community of free nations. Now we are taking steps to defend our people and our environment and the environment of the world."[33] Greening the White House was not only a way in which to create a healthier environment, slow climate change, and generate new jobs, but it was also a useful U.S. geopolitical strategy.

The connection between U.S. foreign policy and U.S. domestic environmental policy is not implicitly a bad one. When Clinton signed the Kyoto Protocol designed to fight global climate change by reducing the greenhouse gas emissions from developed nations by 5 percent from 1990 to 1997 (the U.S. Senate refused to ratify the agreement, fearing it would negatively affect the U.S. economy), he also unapologetically engaged in U.S. environmental geopolitics, though it would be naive to suggest that this exemplifies U.S. hegemony throughout the world. The challenge is that as a world superpower, and as one of the world's worst polluters, the international significance of active U.S. involvement in initiating, advancing, and supporting global environmental policy does not have to arise out of misguided conceptions of U.S. exceptionalism and an exercise in U.S. domination of global political space. The involvement of the U.S. administration in international affairs needs to be more nuanced if the country is to retain its legitimacy as a serious player in favor of social and environmental justice issues. Finally, the rest of the world cannot be perceived as passive participants in this process.

The geopolitical motivations of U.S. environmental policy went completely underground when the forty-third U.S. president, George W. Bush, was sworn into office January 20, 2001. That year, the newly appointed president removed the United States from the Kyoto accord. On February 16, 2005, when 141 nations signed the revised accord, Australia and the United States were the only industrialized nations that refused to sign. At a meeting of 16 carbon-emitting nations in Washington, D.C., in September 2007, Bush explained: "Each nation must decide for itself the right mix of tools and technologies to achieve results that are measurable and environmentally effective."[34]

Bush Junior's preference for voluntary as opposed to mandatory reductions helps explain an apparent contradiction that has left many baffled. Why is the Bush ranch in Crawford, Texas, but not the White House an off-the-grid, ecofriendly home? The surprise and confusion this causes is clearly articulated on the popular *ecorazzi* Web site:

Would the real George W. Bush please stand up? On one hand, we have a President with a terrible environmental record, oil ties, etc. On the other, we have a President with a sustainable home that's off-grid, features geothermal cooling and heating, passive solar, and a grey water system. Whoa? How do you go from installing such features and benefiting from them but not at all pushing them into the mainstream? Does Bush know something we don't? Is he prepared while the rest of us are doomed?[35]

Some of the environmental features architect David Heymann used in the design of Bush's 4,000-square-foot home include: geothermal heating and cooling, an orientation that makes the best use of passive solar heating, recycling of rainwater and household water for irrigation, and the incorporation of native wildflowers and grasses.[36]

Conversely, the green features introduced into the White House during Bush's presidency were minimal and largely a leftover of initiatives Clinton established, namely solar panels providing electricity to the White House grounds and for heating the presidential pool. The decision by the National Park Service to quietly install the panels was a direct result of the Greening of the White House management program's earlier education initiative for staff and maintenance crew regarding the benefits of using environmentally friendly practices and technologies. Yet, why did the use of solar panels receive so little publicity, especially when Bush's environmental record was at an all-time low. As reported in the *Washington Post*, "A spokeswoman for the White House said the administration considered the changes an internal matter that it did not need to publicize."[37] For Bush, sustainability was a matter of individual choice not to be mandated by government or international policy, the exact reason he gave for refusing to sign the Kyoto Protocol. The decision to green his private residence is therefore an extension of his conservative politics, a position that favors individual choice over and above that of governmental restrictions. In short, the private realm of personal choice was rendered intelligible and the greening of the people's house remained invisible and unintelligible all in an effort to legitimate the president's position in favor of *voluntary* as opposed to *mandatory* emissions cuts at Kyoto.

On one level, power can be used productively to address, correct, and transform the history of environmental oppression and abuse. Or, it can be put in the service of a repressive mechanism. Since Carter's time in office, a series of greening and de-greening initiatives have been undertaken throughout the White House complex. The story of the very public incorporation of new sustainable technologies into the White House by both Carter and Clinton, their removal by Reagan, and complete disregard by Bush Junior is political insofar as the discourse of natural life increasingly became both the subject and object of political power. As the White House has been inscribed by the discourse of sustainability the categories of nature and culture, biological life and politics, ecology and transnational corporate capitalism, environmentalism and militarism have been conflated.

5 Green Boots on the Ground

Strategically, a sustainable Army is an innovative Army that can rapidly adapt to future challenges, and an Army that has the support of the Nation it defends, whether in war or peace. In essence, the foundation for such a sustainability ethic is already embedded in the Army core values that inspire us to act with integrity: doing what is right—legally and morally. We are protectors of freedom, and we are warriors with integrity.[1]
—U.S. Army Sustainability Goal

The Cold War left behind a toxic legacy that, depending on which standards were used, was estimated to "cost the U.S. taxpayers between \$330 and \$430 billion to clean up."[2] The initiative to green the U.S. military that President Clinton inaugurated aspired to make a serious dent in a massive task and change the military's attitudes toward the environment. Largely this entailed, and still does, modifying the culture of the military. Primarily, this is achieved by integrating environmental issues and concerns into the military's regular activities, the services it provides, and the products it uses. This translates as managing training ranges and lands with a view to long-term availability, procuring environmentally friendly products that reduce resource consumption, reducing the amount of solid waste and the military's consumption of energy and natural resources, as well as introducing pollution-prevention initiatives (these include recycling of batteries, solvents, fluorescent lights, and nonhazardous wastes).

However, the goal of greening the military fails to distinguish properly between integrating environmental policies into its management systems and the function of the military profession, which Samuel Huntington chillingly described as the "direction, operation, and control of a human organization whose primary function is the application of violence."[3] Exploiting principles of sustainability as part of the military arsenal in effect distorts the fundamental premise of sustainability—working to meet the needs of the present generation without undermining future generations' ability to meet their own needs—if not because the military is ultimately a regressive structure, the very nemesis of civil society and democratic life. If we briefly look to Naomi Klein's "Disaster Capitalism" thesis that demonstrates the complicity between U.S. democracy-building, waging war, and capitalism, then the political goals that the military sets out to realize are ultimately unsustainable at their core. If the capitalist economic engine feeds off of the reconstruction industry in war-torn parts of the world, then any army sustainability goal is a paradox in terms.[4] Clearly, the proposition to transform the culture of the military to be more environmentally friendly and focused on advancing and using principles of sustainability is a cynical exercise and, as argued below, it is used to conceal the fact that the effects of military power are fundamentally unsustainable.

If one rejects that there exists a common ground between the military and sustainable development and recognizes that it is derisive that the military—an organization committed to waging war—is worried about its ecological footprint, then the reality of environmental degradation and human well-being seems very different from the ecogeopolitical conception of environmental security that Braden Allenby defines as "the intersection of environmental and national security considerations at a national policy level."[5] The shuffling of environmental concerns and military values to bring the organization closer to civil society is surprisingly, in many ways, the effect of Left, liberal politics. Bacevich explains, and I would agree with him, that "liberals have grown comfortable with seeing the military establishment itself not as an obstacle to social change but as a venue in which to promote it, pointing the way for the rest of society on matters such as race, gender and sexual orientation."[6] And I would add to the mix of progressive causes Bacevich lists the issue of sustainability.

Contemporary ecogeopolitical discourse combines discourses of ecopolitics and geopolitics. Its arguments primarily fall into two main categories. The first argument wrongfully puts forward a utilitarian line of reasoning: the environment must be protected in order to enhance national and individual security. This position assumes that a sustainable approach to the culture of the military will maximize environmental benefits and hence the security of everyone. The second argument relates to the preservation of U.S. sociopolitical ideals—life, liberty, and the pursuit of happiness—in which the military has mistakenly become the theater in which these are played out. Both positions, which underpin the Clinton military greening initiative that sought to turn environmental issues into a national security concern, overlook the serious implications of applying military-based mechanisms to assess the value of life. For instance, it is wrong to ask the soldier in Guantanamo Bay who is beating a semiconscious prisoner what the value of his victim's life is. The only person who can answer that question is the victim. The same logic applies to how we evaluate the relationship between the military and environmental and social justice issues, for militaristic uses of power are not premised upon a model of collaboration and cooperation; they are oppressive structures of domination. In short, military power does not empower the subject of violence to assert agency in the way that sustainability culture attempts to—in fact, quite the opposite.

The discourse of military power cannot translate seamlessly into a discourse of sustainability. Ultimately, an unbridgeable chasm exists between the fragile truth of civil society and its values, and the military, which is not the same as saying that the military is unnecessary; rather, my point is that the policy to green the military is insincere at best because it conceals the fact that the military's function is to conduct war. And, if anything, the work of the environmental activist or those involved with sustainable development cannot be equated with military systems. This chapter traces how a common ground between the discourse of a U.S. military ethic and that of a sustainable ethic has been constructed, going on to argue that one of the biggest challenges facing sustainability culture is how to reassert their separation.

The military uses the popularity of the discourse and practice of sustainability as a "tool for mission accomplishment" and the maintenance of an asymmetric advantage in respect to perceived threats.[7] My first premise, then, is that the policy to green the U.S. military in an effort to maximize security is merely a smokescreen for U.S. militarism. In 1989, when the Cold War came to a close, the bipolar balance of power set by the standoff between the Soviets and the United States dramatically ended. Accordingly, the singular threat to U.S. security grew elusive. Over time it became apparent that threats to national security were no longer restricted to state actors. Drug traffickers, insurgents, terrorists, organized crime, and environmental degradation all were perceived to pose serious challenges to U.S. national security. Without one dominant threat in place, the meaning of national security became harder to define; meanwhile, the definition of America as the dominant global power went unchallenged.[8] As Clinton's first secretary of state, Warren Christopher declared the world after the fall of the Soviet Union was "a world transformed."[9] The effect of this transformation was the evaporation of politics. As the line between domestic and foreign policy dissolved so too did the political lines delimiting different ideological positions (communism and liberal democracy). In this manner, a limitless principle was anxiously inaugurated as the new mode of political life.

After the Cold War era the crucial question on the minds of those both in the presidential administration and at the Pentagon was: How can the military enhance the security of state when the threat to national security is indiscernible? As Clinton exclaimed in his First Inaugural Address: "Today, a generation raised in the shadows of the Cold War assumes new responsibilities in a world warmed by the sunshine of freedom but threatened still by ancient hatreds and new plagues."[10] One possible answer to the question of who, or what, constituted a threat was presented to Congress in 1995 by the U.S. secretary of defense: "environmental security is now an essential part of the U.S. defense mission and a high priority for the Department of Defense (DoD)."[11] Just five years earlier Gore had published his "Strategic Environment Initiative" (SEI), an ironic reference to Reagan's arms-development program termed the "Strategic Defense Initiative."[12] The proposition was to create and develop environmentally friendly technologies for energy, transportation, manufacturing, construction, agriculture, waste reduction, and recycling. The plan also required wealthy nations to transfer environmentally friendly technologies to economically disadvantaged nations. Although at the time Gore's SEI plan was released, the United States devoted just one-fifth of its budget to energy research and development and two-thirds of it to defense-related spending, strangely Gore was unwilling to finance his environmental policies with defense funds. For this reason, Jon Barnett argues Gore's otherwise reasonable and practical suggestions ended up being reduced to a "set of narrow military and foreign policy responses."[13] As the budget for energy research and development turned into a component of defense spending, the gap between the military and civil life began to close.

Robert Durant provides a detailed narrative of the historical struggle to create a corporate sense of responsibility within the U.S. military toward the health and safety of the public and the environment. The key problem he identifies was how to achieve this without compromising military readiness and, more significantly, how large-scale change could come into effect in a public organization whose culture throughout the Cold War had been defined by sovereignty, secrecy, and sinecure. During this time the Pentagon had argued it was in a better position than national or state regulators to assess and resolve the need for environmental protection within their organization. This position of sovereignty seriously undermined the work of national and state regulators to hold the military accountable to environmental and national resource (ENR) laws. The military decided for reasons of preserving national security which information could be released concerning how it conducted its affairs. Durant tells that the Pentagon went so far as to hire public relations firms to gather statistical evidence to support the claim that environmental stewardship goals were being met. From 1985 to 1995, the DoD claimed it had reduced fuel consumption by 20 percent and the average facility usage by 13.9 percent. In 1998, after having reduced its pesticide use by 50 percent, the EPA awarded the DoD an environmental excellence award.[14]

Under the Clinton administration, the main person responsible for instilling into the culture of the military a strong ethic for environmental stewardship was Sherrie Wasserman Goodman, the deputy undersecretary of defense for environmental security. However, she found serious inaccuracies in how the military reported environmental data and "changes in the definition of what constitutes a site investigation" that together compromised the positive assessment of the military's new environmental record.[15] For all these reasons, Durant states that before military activities could be made to comply with ENR laws, the culture of the organization had to become more transparent and accountable. All this is fine. But it also meant that not only the internal culture underwent a radical overhaul; the role of the military also changed. In other words, the clear and distinct boundary between a civilian organization and the military was blurred as the category of environmental security took center stage.

Over and above the need to introduce environmental and natural resource values into the daily operation of the culture of the military was the more urgent concern of realigning the meaning of sustainable development as part of national security. Clinton's national security advisor, Anthony Lake (1993–1997), claimed the prevailing policy of containment during the Cold

War had become redundant. He advocated a new policy of enlargement, which he understood as expanding democracy and market economics throughout the world. In his "From Containment to Enlargement" speech on September 21, 1993, Lake stressed that America's choices concerning foreign policy would help determine the following:

Whether Americans' real incomes double every 26 years, as they did in the 1960s, or every 36 years, as they did during the late '70s and '80s.

Whether the 25 nations with weapons or mass destruction grow in number or decline.

Whether the next quarter century will see terrorism, which injured or killed more than 2000 Americans during the last quarter century, expand or recede as a threat.

Whether the nations or the world will be more able or less able to address regional disputes, humanitarian needs and the threat of environmental degradation.[16]

Within this picture, environmental and social justice advocates would not only help promote the goals of civil society and strengthen democracy, their work also would contribute to the overall policy of enlargement.

As the fourth part in the strategy of enlargement, sustainable development was seen as key to expanding democracy to the developing world and helping move the planned economies of these countries toward the free market.[17] This narrow version of state-centric environmentalism, working under the umbrella of ecogeopolitical strategizing, was co-opted by the Administration to advance its own economic and military agenda. As Goodman clearly expressed in mid-1993, her commitment to environmental security was to: ensure "responsible environmental performance in defense operations" and to "deter or mitigate impacts of adverse environmental actions [on] international security."[18]

Conflating geopolitical military aspirations with environmental and social justice issues is problematic because it fails to adequately grapple with the shared realities of global environmental degradation and the other issues such as poverty, war, and inequality into which this feeds. Put differently, pollution, stratospheric ozone depletion, clean water, and climate change are a collective problem affecting all forms of life on earth. They are not a selective problem of security, exclusive to any one particular nation. This conundrum at the core of state-centric environmentalism helps explain the nonsensical position of President Bush Junior when he gave his support for international environmental regulations, such as the

Kyoto Protocol, and yet refused to ratify the agreement for reasons that it would compromise U.S. business and security interests. Here he clearly refused to act in the best interests of long-term sustainability goals that would benefit the entire planet in favor of short-term national preferences and interests. This position is a direct result of bringing what are otherwise two irreconcilable discourses—national security and sustainability—together. Namely, if we reduce the issue of sustainability to a problem of national security, then any multilateral international agreement will be assessed on the basis of how it complies with the preferences and interests specific to one particular nation, instead of cooperating to devise pragmatic solutions to a problem collectively shared for the common good of the global community.

Of course, the ecogeopolitical language used in the discussion of defense and international policy is partial. The manner in which ecological degradation is described in militaristic terms is not an objective category separate from the deeper issue of a U.S. culture of militarism. Nor is the political meaning of environmental security constant; the word *sustainability* often used in reference to environmental security is a classic case in point. In the *Road Map to National Security*, a report that set out to define the state of U.S. national security after the collapse of the Soviet Union, the word *sustainability* is invoked twenty-five times in 156 pages.[19] As the meaning of the term is rearranged, it is offered in support of any number of political agendas: America's position as a world leader in science and technology, public interest in biotechnology, overseas combat, U.S. needs, the growth of the U.S. economy, and Democratic Peace.[20] After using *sustainability* in reference to such a broad array of issues, the report suggests broadening the definition of national security as defense: "to include economics, technology, and education for a new age in which novel opportunities and challenges coexist uncertainly with familiar ones."[21] As a result, social justice issues were militarized under the rubric of national defense. Meanwhile, the threat to national security remained elusive.

It is clear that although America was freed of the Soviet threat, the revised notion of what constituted a threat to U.S. national security was inextricably bound up with a new dynamic of uncertainty. The threat of nonstate actors and global climate change both were construed as part of this dynamic. As Senator Sam Nunn, chair of the Armed Forces Committee, had said in 1990:

There is a new and different threat to our national security emerging—the destruction of our environment. The defense establishment has a clear stake in countering this growing threat. I believe that one of our key national security objectives must be to reverse the accelerating pace of environmental destruction around the globe.[22]

Not only does this discourse conceal the antagonism between the effects of military force and the fragile condition of the earth's changing biosphere and the impact this has for all life on earth, it also provides the military with ammunition to continue receiving generous funding from government.

As of February 14, 2007, more than 700,000 active and reserve soldiers were deployed overseas as part of the war on terror.[23] The budget required to finance such large-scale operations is far from being a drop in the ocean. For 2007, the military budget for the DoD was increased $28.5 billion from the previous year, coming in at $439.3 billion.[24] Included within the budget was funding for military personnel; operations and maintenance; procurement; research, development, testing, and evaluation; military construction; family housing; and working capital funds. Disturbingly, these figures did not cover additional funds needed for the war in Iraq and Afghanistan, nor did it cover the budget for Veterans Affairs, Homeland Security, and interest to be paid on past debt-financed defense outlays. These other defense-related expenses saw calculations for 2006 jump from $410.8 billion to $934.9 billion.[25] While the military consumed America's financial resources, the 2006 U.S. Census Bureau reported 36.5 million people in the United States lived in poverty.[26] Viewed in these socioeconomic terms, the military expenditure needed to maintain control over Middle Eastern natural oil resources is simply disgraceful. Furthermore, under such conditions it seems nonsensical to speak of the sustainability goals of the military when we consider the funds used to wage war in Iraq and Afghanistan.

As Bacevich remarks: "Today as never before in their history Americans are enthralled with military power. The global military supremacy that the United States presently enjoys—and is bent on perpetuating—has become central to our national identity."[27] His point is that U.S. society has become increasingly militarized and a few statistics on U.S. defense spending in relation to the rest of the world helps illustrate this point. The *2007 Stockholm International Peace Research Institute (SIPRI) Yearbook* reported:

The world military expenditure in 2006 of \$1158 billion (at constant 2005 prices and exchange rates) represents an increase in real terms of 3.5 percent compared to 2005 and of 37 percent over the 10-year period 1997–2006. The trend in world military expenditure is highly influenced by US military expenditure. In 2006 the \$24 billion real-terms increase in US spending accounted for 62 percent of the \$39 billion total increase in world military expenditure.[28]

All this was happening as "57 million people worldwide" died; "10.5 million of them were children less than five years old. The majority of these children—some 98 percent—were in developing nations."[29] Meanwhile, a byproduct of global poverty of this kind is civil unrest and conflict. For example, the African resource wars were over who controlled the local revenue from the export of valuable resources, such as oil, minerals, diamonds, and timber.[30] Given these facts, it is readily apparent that militarism cannot be equated with environmental and social justice activism.

Generally speaking, environmental and social justice activists aim to build public awareness of issues concerning institutional power, justice, and equality, all the while influencing state actors through a series of intermediary organizations such as Amnesty International, the World Wildlife Fund, or the Sierra Club, which together shape and influence the outcomes of large-scale international agreements. To militarize this process turns the democratic life of collective struggle into a strategic concern of state. It could be argued that the role of the concept of "environmental security" is to enhance the power of democratic government to control democratic life. Take the case of Oregonian environmental and social activist Daniel McGowan, who in 2007 was convicted to seven years in prison for his role in two separate acts of arson in Oregon in 2001.

Working with the Earth Liberation Front, McGowan was involved in the fire at the Superior Lumber Corporation and Jefferson Poplar Farms. In the case of the former, the company had been logging old growth forests in Oregon and the Northwest using helicopter logging, a method infamous for ravaging ecosystems. In the case of the latter, he believed the company was involved with research in the genetic engineering of trees. Explaining why he became involved in the movement, he said:

I had severe reservations about being involved in destroying property, but I felt very strongly about the issues. I felt at the time, we were not getting anywhere with sort of polite protests, very disenchanted with the whole political process . . . The actions were intended to destroy corporate property.[31]

Although nobody was injured, he describes his concern over putting lives at risk, and for this reason he later left the Earth Liberation Front.[32]

When McGowan was indicted in December 2005 for federal charges of arson, property destruction, and conspiracy, he was threatened with serving a life sentence unless he agreed to work for the government as an informant.[33] He refused to participate and eventually pled guilty to some of the charges on the condition that he would remain noncooperative with the state, at which point the government sought a terrorism enhancement for his sentence. If convicted of the charge, McGowan would face a mandatory life sentence with no parole, which would have been the heaviest sentence ever incurred for a victimless act of sabotage in U.S. history. The then U.S. Attorney General, Alberto Gonzalez, described the arrest of eleven environmental activists, of which McGowan was one, in the following manner:

The indictment tells a story of four-and-a-half years of arson, vandalism, violence and destruction claimed to have been executed on behalf of the Animal Liberation Front or Earth Liberation Front, extremist movements known to support acts of domestic terrorism.[34]

In the final verdict for McGowan, the judge ruled one fire was an act of terrorism.

When we consider the destruction wrought on the people and environment of Iraq in the name of national security, the irony of the McGowan terrorist ruling is difficult to dismiss. Who are the real ecoterrorists here? A government who invaded on the basis of faulty intelligence, which has led to the death of approximately 655,000 people in Iraq, a country wholly unconnected with the terrorist attack that incited the war? Or a group of environmental and social activists whose crimes never injured anyone?[35]

The legal discourse quickly slid away from issues concerning a political act to one of national security. As Lauren Regan, executive director of the Eugene-based Civil Liberties Defense Center, explained:

There was no other purpose or reason that this terrorist enhancement should have been applied to ten individuals, ten young people who committed acts of sabotage, which of course are crimes. But the crime of arson and some of the other crimes that these individuals were already charged with carried more than a life sentence. One of Daniel's codefendants was looking at life plus 1150 years for his role in two arsons. But yet the government somehow needed this terrorist enhancement to additionally punish them, if not to label them as terrorists and the resulting chill that would trickle down to the environmental movement, there was absolutely no other legal or other purpose they would have needed this enhancement, other than to go back to Congress and be able to proclaim, look, we have convicted ten terrorists, now give us billions of dollars to continue this fight and give us these tools to legally spy on U.S. citizens, as we know they have done throughout the last several years.[36]

Clearly Regan is arguing that when the disparate concerns of the national security apparatus and democratic life are integrated, the irreverent basis of democracy is quite simply removed.

By trying to define the provocative and unpredictable core of democratic life in terms of a juridico-political form, which Ranciére says can never be identified with democracy proper (the excesses of democratic life that environmental and social justice activism are constituents of), "democratic government" sets out to control the constitutive force of democratic life. The story of democratic government translating McGowan's political act into an act of ecoterrorism is the reason why Ranciére provocatively concludes in the *Hatred of Democracy* that we do not live in democracies.[37]

In exactly the same way as the realist perspective seemed to succeed over the idealism of protesters against the war in Iraq in 2002, it is important to be "consistent in our realism," Ranciére warns.[38] To paraphrase him, as democracy is exported around the world it is not only the useful and positive effects of a constitutional state, elections, and an open media that are brought to people; democracy also means the absence of order. The reason is as Plato once cautioned in the *Republic*, although the idealist premise of political democracy may be to create a government of the people by the people, how this transpires is through the power of the people. Whereas Plato defined the unruly core of democratic life negatively—the wild mob—Ranciére says this is where the strength of democracy lies.

There is no greater strength than people working together for the common good, the outcome of which is never certain—what I call *the struggle for collective life*. The collective process of democratic life (sharing equally the power of ideas and experiences) that Ranciére speaks to, offers an alternative avenue to that of military force, for respect and power. Ultimately, it is the cooperation of people working without guarantees where the primary value of civil society lies. For this reason democracy, in the truest sense of the term, is the government of chance. According to Ranciére's thesis, the paradox of democracy arises out of the following tension: how can a democratic government manage the people without eliminating the excessive and subversive power that the democratic concept of "the people" implies? The paradox arises at all because democracy consists of a great deal more than the right to vote.

Ranciére explains that because Western democratic government has come to mean introducing democracy from outside, "by the armed might of a superpower, meaning not only a State disposing of disproportionate military power, but more generally the power to master democratic disorder," democratic life has been brought under the iron-fist control of an allegedly "good democratic government."[39] President Bush Senior used this kind of reasoning to justify U.S. military control of the Persian Gulf; President Clinton relied upon it to validate his policy of enlargement; President Bush Junior has appealed to it to legitimize the U.S. military's prolonged occupation in Iraq; and McGowan encountered it when he refused to become a government informant and subsequently saw his arrest for property crimes become a matter of national security. Ranciére's point is that the real legitimacy of democratic government is necessarily premised upon the mutual recognition of the subversive power of equality—democratic life of the people—between the people and government all the while acknowledging that in order to remain that way the government cannot afford to seize control of that power. The translation of environmental and social activist concerns into a matter of national security—and, worse still, when these are absorbed by the culture of the U.S. military as a function of military sustainability goals—is nothing other than a deep-seated hatred of democracy.

Some simple questions concerning the logic of harnessing the military to advance the sustainability cause, exposes a fundamental naivety within the ranks of the Democratic Left, who have poorly misconstrued the political importance of maintaining a clear distance between the military and democratic life. It is imperative we ask the hard question: What does sustainability mean for the military in real nuts-and-bolts terms? The opportunity to eat organic food while raping, abusing, and torturing Iraqi inmates at Abu Ghraib? What does the greening of the military mean in terms of the over 34,000 dead innocent Iraqi civilians for the year 2006 alone?[40] Whether or not the U.S. military has used environmentally friendly technologies and equipment for the invasion of Iraq does nothing to change the body count.[41] For the Iraqi civilians who want the U.S. occupation to end, it is irrelevant whether or not the thousands of U.S. military tanks and trucks occupying Iraqi land are hybrids or other fuel-efficient vehicles. More specifically, it is crucial to remember that although we may need the military, we also need to understand the military in terms of it being the most threatening and violent form of strategizing and administration. In this respect, it is important to define the military against the values of civil society and the anarchic democratic life Ranciére speaks of.

In all their eagerness to harness the military to advance humanitarian causes and sustainable principles, Democrats such as Bill Clinton, Gore, and Goodman ultimately failed the Left because they co-opted the democratic life and vitality of the environmental and social justice movement as a form of ecogeopolitical strategizing. The important point is that the military and sustainability are incongruent ideas, so trying to excavate a common ground between them is the ultimate form of camouflage. The military thrives amid ambiguity because it enables disguising the military's regressive structure vis-à-vis progressive civil society. A classic case in point would be the argument over affording the Taliban and al-Qaeda detainees held at Guantanamo Bay the same privileges given to other soldiers under the Geneva Convention (1949), regardless of the fact that they were combatants captured during an armed conflict and that the Taliban fighters in particular were fighting for the armed forces of a signatory party to the Convention at the time of their capture.

President Bush Junior decreed al-Qaeda and Taliban prisoner combatants were not entitled to status as prisoners of war; rather, they were declared to be unlawful combatants. Hence, they would not be "treated legally as prisoners of war" as press secretary Ari Fleischer explained on May 7, 2003.[42] To speak of military sustainability yet ignore the backbone of international humanitarian law is at best farcical. It is the ecogeopolitical categories through which we think military and sustainable principles together that need to become the object of critique for sustainability culture. At best, society tolerates the military, for as an institution it is absolutely anathema to the democratic life Ranciére describes. And it is the impossibility of their congruency that needs to be sustained.

When George Orwell proclaimed in *1984* that "war is peace," he provided the ultimate definition of the military greenwash. Democratic life requires forcing the military to confront what it really does—war—all the while denying it a politically correct exit strategy under the guise of sustainability. It is troubling to think that the military may become the beneficiaries of civil struggles and that they could use these struggles to obscure the violent effects of military might.

Utopia. Photograph by Michael Zaretsky, 2007

II Challenges to Sustainability Culture

6 Trash

"Dirty" Industries: Just between you and me, shouldn't the World Bank be encouraging MORE migration of the dirty industries to the LDCs [Less Developed Countries]?
—Lawrence Summers, former chief economist of the World Bank and Treasury Secretary during the Clinton administration, December 12, 1991[1]

Lead in the paint around the skirting of the house, asbestos tiles in the ceiling at work, pesticides on the vegetables, chemicals in the drinking water, carbon monoxide pumping through the air on the commute, and factories spewing gunk into the atmosphere—these are just a few environmental hazards we might encounter on any given day. The potential dangers are numerous, as are the health problems that arise from overexposure to these everyday toxins, only some of which we can work to avoid. The most we can do immediately to minimize damage to ourselves and the planet is to recycle what we can and, when all else fails, get rid of the rest. What happens, though, when we replace asbestos with ceiling materials that are more environmentally sound; give up old cars that fail emissions tests and buy nice, new, clean hybrids; upgrade to newer, more energy-efficient appliances; replace rechargeable batteries for electric toothbrushes and TV remotes that no longer hold a charge; or even reuse plastic grocery bags for trash bags? The lingering question is where all this old stuff goes. Unfortunately, most of it is shipped to landfills.

Of course, the problem does not end at the landfill; it just takes on a whole new life. As McDonough and Braungart readily acknowledge, the weakness of environmental models that encourage us to follow the four R's—reduce, reuse, recycle, and regulate—lies in the fact that "air, water, and soil do not safely absorb our wastes unless the wastes themselves are completely healthy and biodegradable."[2] Further, trying to find a market for the reuse of wastes may encourage the illusion that we are doing something good for the environment when, in fact, all we are doing is transferring the problem to another place.

Heather Rogers notes that nearly 80 percent of U.S. products are used once and then thrown away.[3] As of 2003, the United States discarded approximately 500 billion pounds of paper, glass, wood, food, metal, clothing, electronics, and other items, a large amount of which ends up in the domestic trash out on the curb once a week.[4] Admittedly, active participation in the recycling options available has dramatically increased; the EPA reports that as of 2007, 32 percent of waste is recycled, which is double the amount from fifteen years ago. Paper, plastic drink bottles, aluminum cans, steel packaging, and appliances are the most commonly recycled materials. Twenty years ago only one curbside recycling program existed, but as of 2005 this had grown to 9,000.[5] Apart from growing consumer concern for the environment popularizing recycling programs, multinational waste organizations have also worked to encourage a recycling culture in the United States. Allied Waste has established collection days for residential households, training programs for business, and pick-up services for larger appliances. The irony is, although corporatizing waste collection may seem to democratize refuse collection by making recycling more accessible to everyone, it has amplified the unequal power relations underpinning the disposal of waste and the resale of recyclable products and materials.

Like most systems that produce a surplus, the collection of refuse has become a multinational business. This trend toward the global trade in trash started to accelerate during the 1970s and 80s, when Browning-Ferris Industries (BFI) and Waste Management Inc. (WMI) were first listed on the New York Stock exchange. Founded in 1969 in Houston, Texas, in the first few years of its operation BFI quickly spread throughout the North American market and by the 1980s it (or its subsidiary companies) was operating in Australia, Canada, the Middle East, Puerto Rico, and Western Europe. By the 1990s when it was bought out by Allied, the second-largest solid waste management company in the United States, annual revenues had reached $5.5 billion.[6]

Rogers narrates the gradual shift in New York's garbage industry from a mafia-led cartel to the new rubbish conglomerates that resulted in the revival of the landfill and the business of exporting trash.[7] In 1993 the EPA passed a series of new initiatives aimed at curbing the negative effects of landfills. The legislation required liners (made of either plastic, clay or composites) be installed under landfills that form like a bathtub underground aimed at preventing leachate from contaminating groundwater, and caps or covers to stop the emission of hazardous gases into the air, along with systems that could effectively monitor both these problems. Rogers points out that this legislation promoted the corporatization of garbage collection; only larger organizations could afford to purchase and upgrade landfills in accordance with the new rules.[8] This resulted in a series of mega-fills that could take thousands of tons of refuse. Further, in order to turn a solid profit, trash collection had to be maximized. Consequently, regional disposal sites were established in poorer states such as Ohio and Michigan. In exchange for receiving trash from cities and other wealthier parts of the country, the states importing trash were rewarded with reduced dumping fees for their own rubbish.

In 1997 the *Michigan Daily* noted the state of Michigan accepted regular household trash from ten U.S. states and Canada, with the state's Department of Environmental Quality reporting that during the 1995–1996 fiscal year the state had disposed of 42.3 million cubic meters of waste in landfills, 14 percent of which came from out of state.[9] In November 2005 the Ohio Environmental Protection Agency issued a report on "2004 Out-of-State Waste." For that year, Ohio imported 3,157,614 tons of waste and exported 1,200,905 tons to five states.[10] At the time, Ohio was only 550,000 tons shy of its 1989 waste import figures, when they had peaked at 3.7 million tons of imported waste—20 percent of all solid waste disposed in America for that year.[11]

The environments of Ohio and Michigan took a beating as their economies crumbled during postindustrialization. As new free trade agreements and the growth of global markets set in during the 1960s, many heavy industry and manufacturing companies in the United States sought cheaper labor in developing countries. A service-oriented industry began to replace the old manufacturing model, putting blue-collar workers out to pasture and making way for a new professional class to emerge.[12] Michael Moore's controversial film *Roger and Me* (1989) vividly documents the debilitating effects of corporate greed and the U.S. democratic myth of equal opportunity and social mobility as he returns to his hometown of Flint, Michigan, to tell the story of how the decline in manufacturing employment bankrupted old established neighborhoods. The motive is clear: to prompt a much needed public debate around the social and economic conditions in the Rustbelt.

In the spirit of a true prankster, Moore sets out to interview General Motors Chairman Roger B. Smith. Taking a well-researched position, Moore narrates the social and cultural impact of GM's policy to outsource labor to Mexico using both facts and his own personal take on the situation:

Well, the million tourists never came to Flint. The Hyatt went bankrupt and was put up for sale, Waterstreet Pavillion saw most of its stores go out of business, and only six months after opening, Autoworld closed due to a lack of visitors. I guess it was like expecting a million people a year to go to New Jersey to Chemicalworld, or a million people going to Valdez, Alaska, for Exxonworld. Some people just don't like to celebrate human tragedy while on vacation.[13]

In true Marxist fashion, the film aspires to raise the consciousness of the audience, provocatively highlighting the unscrupulous side of corporate capitalism. Putting faith in free speech to self-reflexive use, he strategically exercises his constitutional right to speak his mind, in order to show not only the undemocratic side of late capitalism but also the complicity between U.S. media, politics, and corporate life.

If, as the First Amendment of the U.S. Constitution stipulates, a robust democracy depends upon the freedom of speech, then the role of the media as an avenue of free speech in that it exists to provide news that serves the public interest is critical to the health and well-being of any democracy. This is because the media should report on nationally and internationally significant events, in addition to holding politicians, policy makers, corporations, and other influential parties accountable. Moore encourages his viewers to ask the following question: given the power of the media and its extensive influence, why is the precarious situation of Flint's working-class neighborhoods invisible? He makes the content of his film as accessible as possible through comedy, simple language, subjective narration mixed with objective facts, and pop-culture media including cartoons, all in an effort to produce a form of ironic commentary. To query, or worse still dismiss, the factual integrity of the film's message and accordingly place the harsh social realities he narrates in doubt because of the subjective mode of narration he uses (first person) is to remain blind to the deeper politics at work in docutainment of this kind.

The politics of Moore's position does not emerge in its subject matter but rather in its form—not simply his self-reflexive use of free speech, rather how he delivers a potent social and political message by confounding the clarity of the two dominant modes of signification: personal perspective and objective facts. In other words, as the previously cited commentary points to, Moore consciously and strategically mixes subjective and objective positions. Albeit an idiosyncratic documentary style, this blurring of the boundaries between his personal viewpoint and his thoroughly researched factual information constitutes the very abject condition of Flint that Moore narrates. Hence, although at the level of content the film uses a populist visual language, on another level, in confounding the distinctive positions of inside/outside the text, Moore both allows and denies viewers to distance themselves from that content, formally working to confound the clarity of the text to be seamlessly read and interpreted by the viewer.

All in all, the invisibility of the poor in the U.S. Rustbelt is a structural problem of abjection. The body of the impoverished is neither a subject nor an object, rather it is more a threat to the coherency of democratic ideology and corporate power. In this way, Michigan and Ohio become like national corpses, parts of the U.S. landscape that late capitalism has put to death.

Moore is not trying merely to persuade the public of his point of view, for as he combines subjective and objective positions he in effect announces neither are capable of fully signifying the dispirited condition of the Rustbelt. Consequently, as he cuts the power of documentary film down to size, this gesture serves a formal function that enables him to reenact the nonintegrated situation of this part of the country in the popular image of U.S. life and present it as the refuse and waste that such nationalistic imagery tries to dispel. On another level, this depressed and hopeless position the Rustbelt (national waste and yet once the heart of the national economy) holds in the national unconscious is why the filth of wealthier states is allowed to disproportionately soil the land of the poorer states.

Not only is this connection between culture, commerce, and waste a class issue, it also has a racist dimension to it. This is especially transparent when we examine how multinational corporations interact with the economies and societies of the underdeveloped world. Clapp tells how the company Bayer has "admitted that stringent environmental regulations in

Europe have been a main contributor to the movement of their production facilities to Asia."[14] She also cites figures from a United Nations study
showing that half of the multinational corporations surveyed in the Asia-
Pacific region used lower environmental, health, and safety standards than
those they adhered to in developed countries.[15] In turn, when we follow
trade routes of discarded electronics or e-waste exported to the developing
world, another grim picture emerges. As the industry of new technologies
booms hardware quickly becomes outdated: in 2005 a new personal computer entered the market for every discarded one.[16] Now this is not to suggest that we shouldn't update our technologies, however, the design and
disposal of electronic devices needs a radical overhaul if we are to change
the amount of e-waste pumped into the environment, particularly in the
current manner that is disproportionately polluting developing nations.
For instance, by 2002 the world was generating approximately 440 million
tons of hazardous waste, around 10 percent of which is internationally
traded.[17]

According to a report compiled by the Seattle-based Basel Action Network (BAN), approximately 80 percent of North American electronic waste
ends up in Asia.[18] BAN described in one press release the following conditions in Guiyu in Guandong Province, a four-hour drive northeast from
Hong Kong where workers were using nineteenth-century technology to
recycle twenty-first-century waste:

The operations involve men, women and children toiling under primitive conditions, often unaware of the health and environmental hazards involved in operations which include open burning of plastics and wires, riverbank acid works to
extract gold, melting and burning of toxic soldered circuit boards and the cracking
and dumping of toxic lead laden cathode ray tubes. The investigative team witnessed many tons of the E-waste simply being dumped along rivers, in open fields
and irrigation canals in the rice growing area. Already the pollution in Guiyu has
become so devastating that well water is no longer drinkable and thus water has to
be trucked in from 30 kilometers away for the entire population.[19]

One study noted that a used circuit board can be purchased for 13 cents
per kilogram and then resold for 10 cents a kilogram after the metals such
as lead, which sells for US$2.17 a kilogram, copper alloy, US$1.74 a kilogram, and gold have been retrieved.[20]

Many of the developing countries receiving "recyclable" goods from the
developed world do not have the resources or infrastructure needed to
safely recycle or even dispose of these items. Although recycling may seem

an environmentally friendly way to manage waste, when performed improperly the procedures used to extract gold, platinum, and copper from computers, are incredibly toxic for workers. Workers use acid baths to recover metals from circuit boards, sort through toxic materials without any bodily protection, and burn electronic parts on open fires releasing toxic fumes such as dioxins and polycyclic aromatic hydrocarbons. Nisha Thakker explains that if not recycled properly "315 million computers will release 550 million kilograms of lead, 900,000 kilograms of cadmium, and 180,000 kilograms of mercury into the environment."[21] Lead seriously damages the nervous system and cadmium harms kidneys, as does mercury (which also has adverse affects on the brain).

Although the Basel Convention on the Transboundary Movement of Hazardous Waste passed legislation in 1989 to regulate and curb the trade in hazardous waste, especially to poorer nations, the legislation can be enforced only for countries that have ratified the treaty. In 1992, when the legislation could be legally enforced, it had been ratified by only twenty nations, which did not include the United States or Australia. According to the original convention, ratifiers must be committed to the idea that "States should take necessary measures to ensure that the management of hazardous wastes and other wastes including their transboundary movement and disposal is consistent with the protection of human health and the environment whatever the place of disposal."[22] The definition supplied for waste was "substances or objects which are disposed of or are intended to be disposed of or are required to be disposed of by the provisions of national law."[23] So how did so much e-waste end up on the doorsteps of underdeveloped countries?

Article 4.9b of the Basel Convention offers a few clues. It lists some exceptions to the rule, one of which is if "Parties shall take the appropriate measures to ensure that the transboundary movement of hazardous wastes and other wastes only be allowed if . . . The wastes in question are required as a raw material for recycling or recovery industries in the State of import."[24] The loophole comes from how waste is defined in contradistinction to recycling. But there is a crucial difference between recycling nonhazardous as compared to hazardous materials, one that wealthier nations and multinational corporations have chosen to hide their heads in the sand over. McDonough and Braungart hit the nail on the head when they insist a great deal of recycling is merely downcycling, understood to mean

reducing the "quality of a material over time."[25] One of the examples they
provide is the reuse of high-quality steel in automobiles (high-carbon,
high-tensile steel), which is melted down with "other car parts, including
copper from the cables in the car, and the paint and plastics coatings"
which all lower the overall quality of recycled steel.[26]

Apart from the problems of unregulated disposal and downcycling mas-
querading as recycling, questions about the environmental consequences
of waste and recycling are a discourse that is part of a larger phenomenon
of how disciplinary power is exercised. The misuse of international legislation such as the Basel Convention, which is designed to motivate a more
sustainable approach to the management of waste, is telling indeed. It
seems to suggest that sustainability culture is becoming a repressive discourse of which the poor are the subjects. The situation can be viewed
within the epistemological framework of how poor bodies living in underdeveloped areas are colonized. The manipulation of the distinction
between waste, recycling, and downcycling is a practice that determines
whether a subject will count within the norms of a given society; this is
the only way to explain why a company may use certain ethical standards
on its home turf but have no difficulty in suspending these when entering,
for example, China or India.

Whether examining the poor living in the U.S. Rustbelt or the poor
living in underdeveloped nations, how these bodies enter the discourse of
sustainability and environmental law is also part of the problem of how
power institutionally manifests itself. When the "truth" of dumping waste
in underdeveloped nations was exposed and international legislation
against this practice was implemented, workers of the underdeveloped
world were not set free. Instead, this "truth" was used to further legitimate
prevailing power structures driving late capitalism. The management of
the definition also included a system of social regulation that culminated
in the sacrifice of the impoverished body. Power, as used in environmental
discourse, entails the privilege to decide whose communities are polluted
and whose are spared, to define what constitutes a recyclable material on
the soil of a developed country, and in turn to choose which environments
can be sacrificed like common waste. Together these three issues establish
a discourse that is both an effect and instrument of power.

Caught in the trap of being both radically excluded from the clean,
wealthy, coherent national body, and yet also a body who carries the

power to put the clarity of this identity into question, the location of the poor-as-trash within the social fabric takes on a political dimension. Pointing to the abject dimension of trash, Rogers says:

Garbage is the text in which abundance is overwritten by decay and filth: natural substances rot next to art images on discarded plastic packaging; objects of superb design—the spent lightbulb or battery—lie among sanitary napkins and rancid meat scraps. Rubbish is also a border separating the clean and useful from the unclean and dangerous. And trash is the visible interface between everyday life and the deep, often abstract horrors of ecological crisis.[27]

Yet what makes garbage an abject substance is not inherent to the discarded objects themselves. For example, plastic packaging once was used without any consideration for it being a signifier of dirt and filth. It becomes degraded only when it is expelled from such a system of signification and enters a borderline state—a commodity and a noncommodity, clean and dirty, old and new. The crucial point here is, not what the object or substance signifies but the manner in which it resists definition and meaning; put differently, it is ambiguous.[28]

Carrying on from here, as consumer culture introduces into the market an endless stream of objects that are then eagerly consumed, this whole process constitutes the productive drive of capitalist culture. The abject modality of commodity culture demands we forsake those bodies incapable of consuming (poor) and the waste, which no longer carries a surplus value (trash), both of which can longer be signified within the dominant language of consumption. This culture of consumption is radically different from the stewardship culture of bricolage that prevailed throughout the nineteenth century. Bricolage, primarily a women's craft involving the creative reuse of scraps disappeared during the 1950s as commodities became more accessible and affordable.

In her study *Waste and Want*, Susan Strasser aligns the sociological tendency to repair objects to prolong their useful lives with a culture that values durability. When mass consumption became the social and cultural norm a new culture of disposability was inaugurated. More recently with the rise of political environmentalism, recycling became a reality as "Americans became conscious of reuse and waste reduction in everyday lives."[29] Strasser creates historical connections to the countercultural movement of the 1960s and women's housekeeping habits at the turn of the twentieth century. She proposes the shift toward recycling in U.S. cultural practices was less a revolutionary stand against consumerism than a way in which to assert that habits of consumption carried significance beyond individual lives.[30]

For anthropologist Mary Douglas the cultural custom of throwing away points to what she describes as "matter out of place."[31] Meaning, dirt is not intrinsically dirty but instead a relative condition. She writes: "Shoes are not dirty in themselves, but it is dirty to place them on the dining-table; food is not dirty in itself, but it is dirty to leave cooking utensils in the bedroom, or food bespattered on clothing . . . "[32] Similarly sorting and disposing of items does not mean these are inherently useless commodities; rather, the activity is a performance of U.S. identity—contemporary, stylish, clean, competitive, and upwardly mobile. This performance can be described archaeologically by showing the discursive practice of waste at work. An archaeological description of waste is one that follows this transformation of the meaning of waste—bricolage, planned obsolescence, and recycling—then how these function within consumer culture.

Henry Ford's principle of planned obsolescence came into effect during the 1950s. The self-serving character of this approach to business and product design is succinctly epitomized by the 1959 Dodge advertisement whose slogan read: "The old must make way for the new."[33] That is, in order to retain a competitive edge in the market companies need to design and manufacture products with a limited life range. In effect, this type of planned obsolescence was a reactive power in that it referred the affective power of consumption back to an object, albeit one with a short life span.

This shift from stewardship to a disposable culture is interesting because functional failure has been built into the concept of what constitutes "good design." As Strasser notes, by 1981 "Americans held over six million garage sales a year, generating nearly a billion dollars . . . Rather than a sign of a troubled economy, garage sales were a function of affluence, a response to the proliferation of stuff . . . "[34] Continuing on from here, she describes how a countercultural movement started to take hold in the 1960s and 70s as America's youth "celebrated the old-fashioned as subversive."[35] Here, recycling another's waste turned into a cultural politics of sorts, one that set out to critique the dominant consumer model. Playing with the meaning of trash, dissidence lay in the attempt to uncover the arbitrary nature of how we define what is a commodity and what is useless. Further, it exposed the language and grammar used to position these different definitions as inherently violent. The counterculture recycling movement transfers the energy of horror and revulsion toward a joyous and life-affirming mode of production—the positive and affective side of power.

The problem of how to transform the negative concept and abject function of waste is one that reconciliation ecology addresses. Reconciliation ecology, as advanced by Michael L. Rosenzweig, is "the science of inventing, establishing, and maintaining new habitats to conserve species diversity in places where people live, work, or play."[36] For advocates of this position, the focus of preservationists and others who propose we restore an ecological balance to natural habitats insufficiently addresses the hard fact that in many environments the maximum amount of land available for preservation has now been reached. With this in mind, our romantic attachment to the "wilderness" needs some adjustment in order for human beings to be more actively involved in the management of reserves. One way is to introduce species diversity back into all environments, including a city park, a schoolyard, a nature preserve, or even a landfill.

During the latter part of the twentieth century and into the beginning of the twenty-first century the conversion of capped landfills into new urban green spaces became more popular. This is not to suggest that the move to transform a former landfill into green space is without historical precedent, for example Seattle's transformation of its Rainier Dump into Ranier Playfield in 1913, or later in 1935 when part of Seattle's 62-acre Miller Street dump was converted into park land now known as the Washington Park Arboretum. The move toward landfill conversion is indicative of a much larger cultural shift that favors correcting long-standing environmental injustices as well as being a creative response to the economic realities of purchasing urban real estate. In dense urban areas land is scarce and acquiring land in cities is often economically prohibitive, leaving little room for the creation of new public parks to meet the demands of increasing urban densification. Acquiring a landfill and turning it into public space is an attractive option, especially when the seller may even provide maintenance funds for a period of time. So how might the closure and renovation of landfills break away from the prevailing system of abjection and domination? In other words, how can new sustainable ways of life be created? As John Bellamy Foster argues in *Marx's Ecology*, this involves addressing how we interact with the environment.[37]

Foster recognizes that organisms in general do not simply "adapt to their environment; they also affect that environment in various ways, and by affecting it change it."[38] This was the underlying principle used for the redevelopment of the Fresh Kills landfill site on Staten Island. In March 2001 the landfill was closed with one section briefly reopening six months later to deposit the wreckage of 9/11. After workers finished combing through the 1.6 million tons of debris left behind by the collapse of the Twin Towers, the landfill was closed permanently.

Once Fresh Kills was finally closed there still lingered the question of what to do with the 3.5 square mile site—twice the size of New York's Central Park, so large it can be seen from space. Ecologist Steven Handel's solution was to restore biodiversity to the area and create what he describes as "a green island in an urban sea."[39] Instead of simply planting grass over the four mounds to stop erosion or incurring the enormous expense of landscaping the entire area (the estimated cost of maintaining a grass lawn for the first thirty years over the 2,400 acres was $20 million), Handel's suggestion was to start the reconciliation process by planting seven different species of trees and shrubs.[40] The idea seems to resonate with one that McDonough and Braungart put forward in *Cradle-to-Cradle*:

The average lawn is an interesting beast: people plant it, then douse it with artificial fertilizers and dangerous pesticides to make it grow and to keep it uniform—all so that they can hack and mow what they encouraged to grow. And woe to the small yellow flower that rears its head![41]

Handel decided to encourage the small yellow flower to grow.

The concept behind Handel's landscape design for Fresh Kills was simple: attract migratory birds indigenous to the wetlands on which the landfill is built. As the birds fly in from surrounding forests to roost they bring seeds with them, slowly creating new areas of coastal scrub.[42] Handel has successfully skirted the perennial question of anthropocentrism versus ecocentrism, taking instead a coevolutionary approach. Rather than adopt an idealist position positing the question in terms of values (that are the result of how we know the world and the ideas we use to organize it), Handel took a deeper materialist approach to the landfill site by focusing on the historically constituted physical conditions specific to life in that area.

There is a wonderful lesson to be learned from all this: although the concept of abjection may be helpful when trying to understand the power dynamics capitalism depends upon and perpetuates, the notion largely supports an idealist position (reality is not independent of the mind it is simply an extension of our mind) attending primarily to signification or the lack thereof. As such, it fails to address how a subject asserts autonomy and creates alternative frameworks of existence. Rather than appeal to the abstract realm of signification, perhaps we need to return to Marx once more to rediscover the fertile ground of dialectical materialism, with bodies being both the subject and object of change—adapting to the environment and having the power to affect and radically transform it. The key is to create a sustainable connection with the environment by attending to the processes of change implied within that connection. In the context of the discussion here, this would also mean a more dialectical treatment of waste, society, and the environment. The alienation of the poor and the alienation of nature go hand in hand, and both are critical to this conception of life.

In conclusion, if what is repressed comes back to haunt us, the leftovers of excessive consumption take on an abject dimension, generating incomprehensible landscapes that house the familiarity of consumer culture alongside the uglier effects of capitalist modes of production. At the same time, this tension connotes a fundamental ambiguity driving capitalist consumption: the lure and revulsion toward commodities. Capitalism being inherently bound to generating a surplus, a value not assimilated by manufacturing and marketing costs, means that inevitably capitalist modes of production are marked by abject structures and discourses of power. How landfills and the e-waste exported to underdeveloped nations figure in this schema points to a driving problematic of how to effectively reconcile the creative trajectories capital can give rise to with the more impoverished and disenfranchised underbelly conditioning capitalism per se. In this context the landscapes filled with Western e-waste, along with the poor who scavenge a living off of them, suspend and transgress the dominant emphasis given to recycling waste throughout sustainability culture. The more the affective power of sustainability culture is contained as it is represented within a dominating framework, the more environmentalism runs the risk of contributing to dominant apparatuses of power. In so doing, sustainability culture runs the risk of assisting, more than subverting, the institution of subordinating economic, social, and cultural practices.

7 Disaster Relief

One of the effects of militarism and natural disasters is widespread population displacement, out of which usually arise temporary housing initiatives. The figures are alarming: the United Nations High Commission for Refugees (UNHCR) estimated in 2004 that there were 20 million refugees worldwide and an additional 25 million people displaced within the borders of their own country. And these are figures from before the 2004 Asian tsunami, the continuing genocide in Darfur and the war in Iraq—both of which may have begun at the beginning of 2003 but were still raging in 2007—the 2005 Kashmir earthquake, hurricanes Katrina and Rita in 2005, and the Southeast Asian floods in 2007. Additionally, the IPCC predicts by 2050 the rising sea levels, erosion, and agricultural damage incurred by climate change could result in as many as 150 million environmental refugees.[1]

Initial humanitarian disaster-relief efforts focus on quickly meeting people's immediate needs of health care, shelter, food, and water. Only later are transitional settlements established to provide a greater sense of safety and stability to survivors. Transitional settlements share a great deal in common with more conventional urban neighborhoods in respect to population density, economic activity, infrastructure, and social complexity. However, they are radically different to the urban neighborhood insofar as being designed to be a short-term solution (although they often turn into long-term or semipermanent settlements). The fact that such settlements are never intended as permanent structures points to a new social organization that is at once both stable and provisional.

Disaster causes radical changes in social networks and established ways of life, such as breaking or severely undermining family and community networks, redefining or even amplifying gender roles, and weakening the sense of belonging and history and other social values. All these factors suddenly take on a critical function for the sustainability of recovery efforts. Although it is important in the immediate aftermath of a disaster not to lose time by weighing up the pros and cons of different ways to provide relief, turning disaster relief into an exercise in cultural relativism, it still is important for those involved in the relief effort to avoid a one-size-fits-all approach. No single universal characteristic defines how populations experience and respond to disaster and displacement because different cultures and societies have different histories, resources, skills, and experiences informing and shaping their ability to handle crises as they occur. In the context of disaster relief, whereby people have already experienced a violent rupture in their way of life, presuming, for instance, that what is appropriate for a village in India would work for an African-American community of New Orleans, would be simply misguided. That said, although an overriding definition of what we might commonly describe as "population vulnerability" in the aftermath of a disaster may not exist, there is a shared sense of how such vulnerability works. This is the collective aspect of the problem, which is one of how agency and the production of subjectivity can be kept alive in the face of disaster.

Whether the disaster is natural, such as Hurricane Katrina, or man-made, such as war, is irrelevant. In either situation the local economy and the underlying infrastructure that helps a community function and get back on its feet are seriously damaged, not to mention the weakening of cultural identity, such as when significant heritage sites and cultural institutions are destroyed. When combined, these factors negatively affect the sparse reserves of morale a community needs to garner the energy to survive and build their lives anew. In short, disaster is indiscriminate. It debilitates social fabric, its economy, and ecosystems. Responding to the challenges this produces, designers attempt to move beyond the immediate relief effort and create a sustainable community once more, in spite of a settlement being transitional, its buildings and infrastructure only ever intended to be temporary.

The December 26, 2004, Sumatra-Andaman earthquake (commonly referred to as the Indian Ocean Tsunami) measured approximately 9.2 on the Richter scale and crippled some of the poorest communities in the world. Not only did the affected areas suffer terrible economic losses, as rice patties filled with salt water and fishing industries were all but obliterated, survivors also had to rebuild their lives from scratch with few resources. As the world rallied in support, offering funds to help in the relief effort, nongovernmental organization (NGOs) and government agencies worked tirelessly to quickly provide medical relief, emergency supplies, food, and assistance in the establishment of temporary shelters to house the thousands of displaced and homeless. Providing temporary housing under such conditions was certainly a challenge. As Toni Radler, the director of communications for the Christian Children's Fund at the time, explained, the organization was able to provide fishermen with new boats to assist in revitalizing the local economy, but one of the biggest hurdles to the livelihood and survival of the people was shelter.

In some of the hardest-hit areas along the Indian coastline, on average 150 shelters per village were established. The conditions inside were often unbearable. Families were living in single-room corrugated tin shelters set on concrete slabs, measuring approximately 8 by 12 feet. They ate, cooked, slept, and took shelter in what was basically a windowless box. With an absence of ventilation along with soaring temperatures reaching 100 degrees Fahrenheit during the day, it is unsurprising that children reportedly suffered from jaundice, fever, sweating, and respiratory illnesses as a result of heat exposure. Trying to alleviate the heat effect inside their shelters, people began to cover the tin with dried palm leaves.

Remarkably, by modifying their shelters the villagers started to resignify the design conditions given to them; in so doing they created new discursive realities that prompted designers to pay closer attention to the connection between shelter, violence, and power. The use of palm leaves to help cool the interior exposed the debilitating and inhumane characteristics of the provisional shelters, resignifying the shelters as the very antithesis of "aid" and "relief," terms which usually denotes a benign and caring response. The repressive function of the shelters produced a ground on which subjective agency began to be asserted. This is not to suggest that designers need to create oppressive conditions in order to trigger a sense of agency; it does imply, though, that agency is not simply a matter of choice or intention. Rather, it is an effect of social discourse. From this standpoint, shelter modification constitutes a form of politics, because the villagers repeated the minimum requirements of what defines a shelter, as classified by the UNHCR, and in so doing they amended this definition by exposing the oppressive living conditions such guidelines can produce.

In reference to providing shelter for Internally Displaced Persons (IDPs) the UNHCR *Handbook for Emergencies* stipulates:

Shelter must at a minimum, provide protection from the elements, space to live and store belongings, privacy and emotional security. Shelter is likely to be one of the most important determinations of general living conditions and is often one of the largest items of nonrecurring expenditure. While the basic need for shelter is similar in most emergencies, such considerations as the kind of housing needed, what materials and design are used, who constructs the housing and how long it must last will differ significantly in each situation.[2]

Generally speaking, the function of a shelter according to this document is to ward off exposure to natural elements and provide security. Shelter as such helps curb incidences of poor health and disease.

Yet, if we take the tin box temporary housing schemes established in parts of India after the tsunami as our point of critical departure for a moment, then the social discourse defining the minimum requirements as laid out in the *Handbook for Emergencies* no longer seem so neutral and benign; that is, it has a sociopolitical undercurrent to it. The subject of design asserts their agency through social discourse, like the simple modifications to the tin boxes. Deriving agency from the conditions of transitional shelters or settlements (designed to last five to ten years) entails establishing an opportunity that is culturally constituted. This means aligning the design process with a primary set of material and historical conditions that together prompt sociality: social patterns, available resources, reliable materials, ventilation, durability, and traditional typologies. The first point to address, therefore, is how social conditions construct subjects and what kind of subjects they produce.

Strictly speaking, IDPs do not exist as such; they are the subjects of social and political practice (NGOs, government, health officials, and international aid schemes).[3] That is, they are a political category, one that is constructed through the inscription of positive and negative traits. This inscription of the IDP body through a set of minimum shelter requirements poses the question of their subjectivity as subjected to these institutionalized norms of what constitutes privacy and the public realm, the sanitized and unhealthy body, as well as permanent and transitory ways of life. In addition the subject produced by this inscription process is temporally discontinuous in that the tsunami cut through the narrative continuity that comes from having been someone's father, husband, son, or brother, only to then be reinscribed as an "IDP." The design and development of a sustainable shelter could use this inscription differently, aspiring to generate opportunities for continuity to reemerge.

A case in point would be the work of ARTES, a Chennai-based group of architects who have been involved with building shelters and settlements in response to Indian areas affected by the tsunami. According to Nandan, who works with ARTES, it is important to make modifications to standard shelter layout with women and children in mind. For instance, providing children with their own space and private areas for women. Or, when shelters are arranged in a geometric grid stacked in rows of ten with narrow eight-by-ten-foot gaps between rows, villagers are unable to keep the animals they otherwise would have wandering around in the spaces outside of their homes and on which they depend as part of subsistence farming efforts. Small compact spaces where people cook inside create enormous problems from the accumulation of smoke. Using outdoor canopies or hut extensions, which draw upon traditional architectural typologies that often situate kitchens outside of homes, improves indoor air quality and prevents the rise of skin disease among children that results from exposure to fumes.

Further, the introduction of toilets and sanitation may be an entirely foreign design solution for some communities, many of whom Nandan reports were unfamiliar with the idea of using a bathroom or toilet. He explains:

It is not really an issue of just providing the amenities. If one goes through the region, one can find abandoned bathrooms, abandoned toilets that are not used. Not the least because they are uncomfortable while using them in a particular manner. Now when we did a detailed study of this in this community for several months the major issue was privacy. And another major issue was the fact that there was not enough light in bathrooms. So, even if they have to use it after sunset or before sunrise there was a certain fear even to walk to those places and which is why they are abandoned. We thought of the idea of using solar lighting in these bathrooms. We tested it out in certain areas in Cuddalore and found out it was a simple intervention that allowed the community to use these bathrooms.[4]

What Nandan suggests is there needs to be a greater sense of continuity established between predisaster and transitional shelter design.

However, adding to this the suggestion is that when new sanitary solutions, and by implication physical changes, are made design turns into a problem of "how" to establish a different social activity in the community:

The toilet and sanitation problem is really a complex one. I don't see it as an issue of only physical intervention or design solutions. It is really a social issue where in the context of a trauma we are also attempting to bring in new different, if the word is better, standards of sanitation. Because we found most of the communities are not accustomed or used to the idea of bathrooms or toilets. In that context, bringing in decisions, sometimes, and the manner in which it is brought, that was also critical.[5]

What this position does not consider are the political and cultural implications of introducing new sanitation habits into a community. As chapter six discusses, the topic of cleanliness and dirtiness is in itself a cultural construction. When we perceive a person, or an entire culture different from our own, as dirty we are often projecting our own social values and the repressions that produce them onto the other culture. These observations reach into the realm of psychoanalytic theory, which claims socialization manifests itself through attention to bodily cleanliness. The sense of disgust and embarrassment associated with everyday bodily functions and odor mobilize the repression of bodily secretions, energies, and impulses in the production of social value.

A supposedly simple and benign design initiative such as toilets and bathrooms is not value-neutral, and depending on the context it can operate as a coercive mechanism. Design now takes on a political role as disaster relief produces the "Otherness" of the abject body of the survivor of disaster, all the while reinscribing it within middle-class standards of morality. The concern here is that teaching new social habits and cultural practices through disaster relief may gloss over a deeper racist revulsion against "unclean" or "untouchable" social groups. My point is that in order to be sustainable, design in the wake of disaster cannot predetermine the life of individual and collective bodies with material and social patterns that are largely external to the life of a given community. It cannot become an extension of style and good taste and at the same time propound to revive the messy sensibility of everyday life. Sustainable design cannot afford to be arbitrary in the face of traditional social patterns and the manner in which these inform human settlements; it therefore must engage with the more spontaneous and expressive life of bodies and the independent social patterns these produce.

As Charles Correa, Paul Oliver, and Geoffrey Payne have made explicit, to be sustainable shelters must work with traditional social patterns and ways of life.[6] The involvement of the community in building their own shelters is critical. When nonlocal skills and traditional building methods are used, involving local villagers in both the design process and construction of their buildings develops a sense of ownership and trains the villagers in new skills that later can be used to maintain the shelters. This increases the possibility that shelters will last longer. There is always a large skill base in place that designers can work with, and doing so avoids adopting a condescending approach to the subjects of design by constructing the survivor as vulnerable and in need of saving. By implication, such a construction allows the designer to emerge as a more powerful entity than the subject of design. In this context design is clearly approached in terms of establishing temporal continuity through social patterning more than being just a spatial problem of how to physically intervene. According to this perspective, the social patterns that produce subjects over time provide an armature for design.

How this temporal disjunction between the pre- and postdisaster subject is located within a much larger discussion around the increasing threats global climate change poses. Without doubt, the issue of global climate change started to enter the social imaginary at the beginning of the twenty-first century. The more it entered circulation the more sustainability became a normative concern for how governments would plan their economies and put new technologies to work, and how they, in collaboration with large international aid agencies, would respond to disaster.

Figures on the increasing number of natural disasters worldwide are certainly alarming and they prompt us to seriously consider not only measures necessary to mitigate the process and effects of global climate change, but also to address how we as human beings will turn these constraints into creative opportunities. As former UN Secretary General Kofi Annan reported in October 2006:

Over the last decade, disasters triggered by natural hazards have claimed more than 600,000 lives and affected more than 2.4 billion people, the majority of them in developing countries. Years of development gains have been lost, deepening poverty for millions and leaving them even more exposed to future natural hazards. Now more than ever, we must accelerate our efforts to reduce vulnerability.[7]

Adding to this, on August 9, 2007, UN Deputy Emergency Relief Coordinator, Margaret Wahlström, reported that between 55 and 65 percent of all annual global disasters are weather-related. For the period 2004 to 2006, floods increased from 60 to 100 floods annually; the overall number of weather-related emergencies rose from 200 to 400; and approximately 500 million people are being negatively affected by these trends on an annual basis.[8]

These statistics not only inform design initiatives for sustainable development aimed at providing a realistic approach to the environment and local climate conditions, they are also used by government and international organizations when representing the interests of what they perceive to be social groups especially vulnerable to climate change and the demise of natural resources. Here the social issue of climate change turns into a political category, and more ominously there is a whole new world order emerging out of the impending problem of climate change. For example, after the tsunami, the Indian government engaged in a series of preparedness planning initiatives aimed at minimizing unnecessary exposure to natural hazards in response to predictions alerting them to rises in sea level and increased flooding. Just to quickly put things in perspective, in the summer of 2002 the Greenland and Arctic ice caps shrank 400,000 square miles (the largest decrease on record).[9] Now in Bangladesh alone, more than 10 million people live within three feet of sea level.[10] With the Intergovernmental Panel on Climate Change predicting a 15 to 95 centimeter rise in sea levels, this would put approximately one fifth of Bangladesh under water permanently, leaving nearly 35 million people homeless.[11] The 1960s saw approximately 7 million people worldwide affected by annual floods; by 2004 the figure rose to 150 million.[12] Given the mounting threat of flooding in India and the social and economic hardships such floods produce, the Indian government's response appears the most responsible course of action to take. Yet for tsunami survivors, the appropriateness of the response was not so self-evident.

Reluctant to let surviving villagers rebuild on the same oceanfront sites they inhabited prior to the tsunami, the Indian government announced it would enforce protective setbacks anywhere between 200 and 500 meters. As the villagers rely upon manpower alone to launch their heavy boats into the ocean, it would be virtually impossible for them to drag their boats that far to the shoreline each day. The economic consequences of the decision are dire, not to mention that the timing was unreasonable. Villagers would be prevented from putting the few remaining assets they had (natural resources and skills) to productive use at a time when their other physical assets, such as homes, belongings, and psychological strength (through loss of friends and family and the weight of traumatic memories) had already been compromised. The villagers appealed to the government, explaining they would not be able to continue fishing if they were relocated too far from the shoreline.

In this case, global climate change became both a political category (used by the government) and a universal representation (absolute truth), both with what Theodor Adorno might have described as a "violent and repressive character."[13] As Adorno maintained, the authentic work of art is the last bastion of reason in a rationalized world because it is the antithesis of a repressive logic subsuming life under a rubric of identity. Sustainable design, like the authentic artwork Adorno speaks of, can provide an alternative rationality, one that synthesizes difference without integrating the particular and nonidentical into a normative value system.[14] It does this by interrogating its own function and meaning in relation to instrumental reason. The sociopolitical discourse of global climate change used by the Indian government constituted a repressive logic because it re-represented the interests of the villagers on the government's terms, doing violence to their rich cultural heritage. And yet its own position was an extension of the international discourse on global climate change. Overshadowing the Indian government's decision were the power structures defining the global arena (developing and developed nations).

There are serious cultural implications to the Indian government's decision to permanently relocate the tsunami survivors away from the ocean; traditional fishing skills that were handed down from generation to generation would be lost and heritage ties to the land would be broken. Unable to fish, the opportunity to earn a living for their families would be limited, increasing stress levels within individual families and in the community as a whole. Understandably, being excluded from the decision-making process and establishment of future development objectives made many villagers suspicious of the government's intentions, fearing that the tsunami was being used as an excuse to remove them from valuable real estate to make way for oceanfront development.

Given the complex discourse framing and informing this situation, what might a sustainable response entail? Design can set out to embrace a trans-subjective approach that aims to empower and promote subjective agency. For instance, why not place the new settlements on the waterfront and design the village in such a way that the buildings are conceived of as part of a much larger ecological process? Under such a design, as sea levels rose and the time came for villagers to relocate, they would simply dismantle their homes and rebuild them farther inland using the same materials. The work Oxfam carried out in Sri Lanka after the tsunami offers an opportunity to consider this idea.

After the tsunami, many people in Tangalle, Sri Lanka, found shelter with other family members or friends. However, a group of seventeen families did not. The families wanted to remain in the vicinity of the rest of their community, yet no land was available. Eventually, a transitional settlement was established in the middle of the village, on the site of the children's playground. For Oxfam, the settlement site was critical in maintaining ties with the rest of the village and allowed easy access to the infrastructure of the village. The result was a series of affordable units that each cost only US$580, a "safe shelter that would enable" the villagers to "store their belongings securely and would be spacious and cool enough to carry out everyday tasks, such as mending nets or drying fish."[15] The overriding design principle was to produce housing that could be dismantled and later reused in constructing permanent housing. Timber joints were bolted together, and in place of a cement slab removable cement tiles were used. Finally, the design was the direct outcome of group discussions with the villagers, local government, and Oxfam.

The policy to permanently relocate Indian villagers away from the shore made no room for the idea of design-as-social-discourse, a mode of design practice that the above Oxfam example works so hard to instigate. In this way, the notion of agency that occurs at the limits of design language, whereby those limits are not always negative constraints but possibilities for creative change, is oppressed by the very terms used to articulate the meaning and value of safety, protection, and well-being. Embedded within the Indian government's decision was official language, no different from the minimum shelter classifications supplied by UNHCR, that objectified the material effects of disaster and how these impact a given subject or collective. For example, the government did not adequately recognize the strength of the population's desire to return to and rebuild upon sites of disaster despite the rubble and debris. Indeed people returned in spite of the possibility of future natural disasters ravaging the area. It is therefore imperative to recognize this longing to return to the shore not simply as a melancholic gesture but also as a yearning to recover the power that emerges out of very deep connections between history, landscape, and the body. This connection is largely affective, and the desire to return is intrinsically related to a deeper will to survive that emanates from the utopian potential implied within the debris.

For the villagers, to rebuild on the shoreline was an affective investment not in the past—as melancholic longing or the concept of nostalgia implies—but in the future. The past was not reified, as the theory of melancholic yearning presupposes, rather it reawakened the possibility to hope once more.[16] As such, the debris, the past, and the will to return meant that the oceanfront site had become a political ground through which agency was produced. Looking to the shoreline, the villagers saw in the wreckage the potential to recuperate the past as a way of surviving against all odds. Indeed, the specificity of this area of land came from all the traces of history embedded within it. In this light, the concern over the land being used for waterfront real estate development was symptomatic of a deeper worry that the area would be reified and its affective dimension compromised. The longing to return was the very antithesis of such reification and the subsequent commodification of the site that this would prompt. In a nutshell, the rubble provided the material ground out of which a sense of hope in the future could emerge once more.

Sustainable design in the wake of disaster and the subsequent development and planning to resituate survivors in more permanent settlements needs to ask the question of how design can negotiate different material histories.[17] The language of design and its organization has a materiality, one that extends beyond just the materials used, the durability of the structure, and the demographic behind the labor used to construct these. That materiality consists of historical ties and the affects such ties produce throughout the collective body.

A collective is not an isolated entity; it is constituted in connection with other singular and collective entities. For instance, without the support of local populations and government the site location of transitional settlements can be hindered, if not stopped altogether. Access to land and issues of settlement placement can be a source of potential conflict as Corsellis and Vitale's study on transitional settlements found, because "when local communities have traditional or customary rights to land that a national government has made available for transitional settlement" sometimes there is the fear of land depreciation in areas where transitional shelters are established, or concerns that displaced populations bring with them destabilizing social, environmental, and political implications.[18] These issues all presuppose that the form of a transitional settlement is predominantly an isolated unit introduced like a foreign body into a preexisting community. However, as the Oxfam example in Sri Lanka shows, when the settlement is considered in connection with the surrounding community it strengthens the sustainability of both the transitional and permanent community, as each works in synch with the other.

Regardless of how good its intentions are, no organization, government, or individual can fully represent or define the needs of another subject, let alone one who occupies a disadvantaged position. Any definition a large organization with enormous power and a vast bureaucracy produces will tend to be implicated in the very social conditions that enable it to maintain an eminent position in the international political arena. In response, clarity on how a sustainable design practice might work in the context of such institutional and structural constraints is imperative, and there are two key issues design can engage with to assist in this regard. First, sustainable design initiatives must respond creatively to these structural limitations and the subjects they produce. Second, such initiatives arise out of these selfsame constraints, which in turn become the conditions out of which subjective agency can emerge. All in all, design must involve itself with the social conditions disaster creates and how this distorts pregiven power relations. Some of these include: broken social networks, overcrowded living conditions, poor shelter quality, distance from employment, loss of productive assets, illness, and unfamiliar circumstances. The overall effect is a continued sense of economic and psychological instability, which when left unchecked can compound the violence of preexisting social norms.

In this regard, Oxfam International's Briefing Report on "The Tsunami's Impact on Women" is most informative:

Experience of natural disasters in a wide range of contexts shows that events of this type can weaken the status of women and girls and their ability to negotiate both within and outside the family. The loss of assets, homes, and family members all contribute to increased gender inequality.[19]

The study shows that the tsunami killed four times as many women as compared to men. For example, out of 676 survivors from four villages in the Aceh Besar district of Indonesia, only 189 were women. And in the four villages in the North Aceh district, 77 percent of those killed by the tsunami were female.[20] The reasons behind the uneven number of male survivors as compared to females are varied and differed from region to region: women stayed behind to look after children and other relatives, men were stronger and able to swim, women were on the shorefront waiting for the fishing boats to return so that they could clean and sell their fish at the local market, and in Sri Lanka in the Batticoloa District this was the time of day when women took a bath in the ocean.

The Oxfam report asks: Given the existence of extreme demographic changes, how safe are the women in the crowded camps? Inevitably the disproportionate survival rate of men in areas hit by the tsunami means that traditional gender roles and family structures would significantly change. Men would have to learn how to care for children and take on domestic duties in a traditional cultural context that views such work as specifically feminine. How can design respond to this situation? Design must somehow allow the marginalized to speak for themselves and on their own terms. This first entails identifying how a disaster disproportionately affects minority groups and then how design can create a common ground from which to act.[21] A case in point would be the situation of African-American women in the New Orleans and Gulf Coast regions after Hurricane Katrina. Both the Institute for Women's Policy Research and the Women's Funding Network discovered that prior to Hurricane Katrina women were some of the most disadvantaged members in the New Orleans metropolitan area. Statistics show women were the heads of 56 percent of households with children in New Orleans. In storm-affected areas, prior to the storm women were more "likely than men to live in poverty, to raise children on their own, and to hold low-paying jobs."[22] Unsurprisingly then, the "scale of the disaster was magnified by long-standing inequalities that placed disproportionate harm and hardship in the path of low-income families, black communities, and female-headed households."[23]

In July 2006 the unemployment rate for all working-age people displaced by the hurricanes was estimated at 11.9 percent.[24] After the storms, the size of this population decreased 40 percent while the Black population fell 65 percent. As a result, New Orleans became "more white, less Black, higher income" and these people were "more likely to own their own home."[25] What is especially interesting is that a large proportion of those who disappeared included poor female households with families. These families made up only 18.3 percent of the metropolitan statistical area (MSA) after Hurricane Katrina, as compared to what had previously been 35 percent.[26] Meanwhile, after the storm the median income for men in New Orleans rose to $43,055 while women's fell to $28,932. Added to this, only ten federally subsidized child care facilities had reopened by late 2006; these had to serve approximately 225,000 people.[27] At the same time, approximately 38,000 families were living in trailers, each measuring 240 square feet, and justifiably more than half the mothers and female caregivers living in these conditions showed signs of depression or anxiety, with less than one in five having sought counseling.[28]

After the widespread media coverage of Hurricane Katrina it was clear that the poorest neighborhoods, the majority of them African-American, had born the brunt of the disaster. The media did not cover well the manner in which African-American women were disproportionately affected, and the reason lay in social and economic inequalities that existed prior to the storms. Prior to the storms white women tended to work as lawyers, post-secondary teachers, and waitresses. Meanwhile, African-American women were more commonly employed as maids, health aids, and cooks. In addition, there were racist distinctions underscoring white-collar professions such as secretaries and administrative assistants where both African-American and white women were employed. Even more problematic, on average white women annually earned $9,000 more than African-American women working in the same profession.[29] Unsurprisingly, when the median income for all working female African-Americans in New Orleans was compared with all white women, there existed an overwhelming disparity between the two groups. The former earned on average $19,951 compared to $36,445 for white women.[30] Yet the aforementioned statistics do not articulate a complete picture and when used in the wrong way can be deployed in a positivist fashion, defining one particular social group as vulnerable, weak, and defenseless. In defiance of such a view, Derrick Johnson, president of the Mississippi State Conference on the NAACP clarified: "African American women are the backbone of any movement in our communities, in the churches, and as individuals doing the work on the ground, often without the titles."[31]

Many women who wanted to return and participate in the rebuilding effort could not afford to as housing costs increased.[32] Given the limited housing stock available and the unaffordable nature of it, federal government could invest in repairing, where possible, houses, apartments, and public housing that were not ruined by the storm, as well as develop new affordable housing. This is in place of approving "$31 million in contracts to demolish 4,500 units of public housing" that had "mostly escaped Katrina's destructive path" and in their stead building a mixed-income development with approximately 1,000 market-rate and 750 affordable units, as reported in *Architectural Record*. All this was planned to take place amid a housing-shortage crisis for working families from the Lower Ninth Ward in New Orleans.[33]

Under these circumstances some of the most effective forms of assistance include affordable housing, schools, health care services, and child care. Although the storms accentuated deep racial and gender inequalities historically specific to the South, they also provided an incredible opportunity to address and begin to correct them in much the same way as the current Make It Right (MIR) project, which was initiated by activist actor Brad Pitt. MIR aims to build 150 affordable, ecologically friendly homes in the devastated low-income neighborhood of the Lower Ninth Ward. The core architecture team includes William McDonough and Partners, the nonprofit Cherokee Gives Back Foundation, and Graft architecture, which is working with local neighborhood-led coalitions of former residents from the ward and other not-for-profit organizations to rebuild the area. The design team is experimenting with typologies that were common throughout the neighborhood—Shotgun, Creole Cottage, and Camelback housing types—in an effort to revitalize the cultural specificity of the area with a view to creating homes that use the following four criteria as their benchmark: safety, affordability, sustainability, and high-quality design. Most admirable about the project is the manner in which it reinscribes the triad of gender, race, and class without adopting one subject position in particular. For example, the needs-based financial package offered to returning residents ensures that not only is their home affordable but the cost can be supplemented with MIR loans. In addition, residents are involved with selecting a design that suits both their aesthetic preferences and lifestyle needs. Further, the area chosen for rebuilding offers residents access to schools, public services, health and safety, infrastructure, and transportation, all of which as I have argued are crucial to reviving the social, economic, and cultural life of the area. For these reasons, MIR embraces a multiplicity of subject positions and demonstrates that, at its most political, sustainability culture demands we struggle to find some kind of common ground from which to act collectively.

During the 1970s, approximately 110 natural disasters affected 740 million people, as compared to 2,935 between 1993 and 2002, affecting 2.5 billion people. During this time the cost of damages increased fivefold, coming to US$655 billion.[34] The changes in extreme weather patterns that the world is currently experiencing are in all likelihood only the tip of the iceberg. Scientist Tim Flannery reports that there are "disturbing signs that hurricanes are becoming more frequent in North America," and the statistics he provides in support of this are hair-raising.[35] Over the course of four years (1996–1999) the United States experienced more than twice the number of hurricanes it typically endured on an annual basis over the course of the rest of the twentieth century. Flannery explains that for "every 18 degrees Fahrenheit increase in its temperature, the amount of water vapor that the air can hold doubles" dramatically increasing the amount of hurricane fuel.[36]

The Earth's average temperature over the last 10,000 years has been set at around 57 degrees Fahrenheit, but as the Earth's temperature rises, so too will the incidence of hurricanes. The rising sea levels caused by the increases in temperature are the result of not only the melting glaciers and ice caps but also the result of oceans expanding—thermal expansion—and over the next 500 years thermal expansion alone is estimated to cause sea levels to rise by 20 to 80 inches.[37] Flannery adds as an important caveat to these predictions that if the West Antarctic ice sheet ever detaches from the sea floor, by 2100 this would add between 6 and 20 inches to the sea-level rise. What is more, this would produce a ripple effect as other glaciers feeding into the West Antarctic ice sheet could probably add 0.9 million cubic miles of sea and glacial ice to the equation, making global seal levels rise anywhere between 20 to 23 feet.[38]

Everyone will be affected by the Earth's rising temperatures; however, as this chapter highlights this will not impact everyone equally. It is here where culture has an important role to play, because as climate change dramatically changes people's lives, the more vulnerable members of the community will be disproportionately affected. In order to remain alert to the specific needs of marginalized groups, a healthy sense of common history is the only platform from which to articulate such difference. An astute awareness of issues of class, race, gender, and the like is important, and it also is important to remain mindful of how these operate transversally. Cultural practices may be unable to fully represent difference but they can reinscribe the manner in which culture intersects and transforms difference, all the while destabilizing its own position of privilege in the process. Following on from here, the sustainability of any given solution in the context of a disaster zone lies less in how authentically it represents those who have been marginalized by an event than it does with producing a positive use value from the debris disaster leaves behind.

8 Slums

In *Planet of the Slums* geographer Mike Davis describes a new urban order, one defined by the rise of megacities with more than 8 million inhabitants, hypercities with populations over 12 million, and increasing disparity between cities of different sizes and economic specialization. If slum settlements are home to approximately one billion people and the developing world is projected to absorb 95 percent of the world's total population growth, then the task facing urban developers, planners, policy makers, and designers is daunting indeed.[1] To date little planning and development has been undertaken to accommodate the growth rate of urban populations in many parts of the world, resulting in slum growth outpacing urban growth.[2]

If in the past the model has been either to eliminate or to relocate slums to high-rise public housing enclaves on the outskirts of cities, where they remain unseen and unheard, then the model of urban integration, an idea that the Favela-Bairro program in Rio de Janeiro became a showcase for, could be said to adopt an entirely different approach to the question of how to implement a sustainable focus on slum development. The aim of Favela-Bairro was to address the housing and support needs of approximately 500 to 2,500 households. The objectives were to "improve the living conditions of the urban poor, and in doing so the program contained a rather wide mix of different social infrastructure, land tenure, and social development components."[3] With a budget of $3,500 to be spent on each family (a figure that later increased to $4,000) it was estimated that Favela-Bairro would affect nearly 60 percent of the overall favela population.[4] Phase one focused on improving health and public services; phase two directed its attentions to social factors such as education, training programs, and the establishment of nurseries.

The program was hailed by many as a beacon of success in the growing field of urban design, with its architect, Jorge Mario Jáuregui, receiving the Sixth Veronica Rudge Green Prize in Urban Design and the Grand Prize at the Fourth International Architecture Biennial in São Paolo, Brazil. However, more recently Favela-Bairro has come under heavy criticism from UN-HABITAT. The organization explained:

despite a long history and commitment to improving the lives of slum dwellers, Brazil has been unsuccessful in improving the lives of the poorest of the urban poor; inequality and chronic poverty are on the rise, and perceptions about *favela* (slum) dwellers have not changed . . . Clearly, there is a need to change people's perceptions about slum dwellers and institute reforms that go beyond slum upgrading and regularization in order to safeguard poor people's livelihood.[5]

This raises an important point common to all sustainable design initiatives: regardless of the content of a project—in the case of Favela-Bairro, urban integration through infrastructure and services plus ensuring housing rights and provisions for environmental sanitation—if the negative stereotypes and perceptions of slum dwellers persist, no amount of investment in these areas will be sustainable. If design is to develop urban slums sustainably, it must somehow engage with the historicity producing such perceptions, and how that history might pose both the limits to and conditions for autonomy.

Rio's favelas have a long history extending back to the latter part of the nineteenth century when soldiers of the Fourth Brazilian Expeditionary forces returned to the city after fighting separatist groups in the State of Bahia in the northeast of Brazil. Allowed to temporarily settle in Santo Antonio and the Providencia hills after being denied remuneration for their service in the form of property rights, they built shantytowns out of makeshift materials. The term *favela* arose because the area settled resembled the informal urban pattern of *Morro da Favela* in *Canudos* (where the soldiers had once fought the separatists). During the early part of the twentieth century through to the mid-1950s the dominant perception was that favelas threatened the decorum and order of the formal city. Hence, the primary objective behind policies such as *Reforma Passos*, *Plano Agache*, and the 1937 Building Code was to eradicate the shantytowns. This often took place without any consideration for providing alternative housing. For example, *Reforma Passos* displaced more than 3,000 households, replacing these with a mere 120 homes.[6]

On April 1, 1964, Brazil entered a period of military dictatorship that lasted until 1984. During this time, as Janice Perlman has pointed out, the definition of the favela as a social problem persisted, resulting in more than 101,000 people being forced out of their homes. They were subsequently relocated into public housing projects that tended to be "several hours and costly bus rides away from the previous sites of life and work."[7] Regardless of a series of aggressive elimination policies, over time the percentage of Rio's population living in informal settlements grew from 7 percent in 1950 to nearly 14 percent in 1990.[8]

In an effort to combat urban poverty through social and physical assimilation, the Municipal Government of Rio with funding from the Inter-American Development Bank (two investment loans, in 1995 and 2000) set out to integrate thirty-eight medium-sized favelas into the broader urban fabric. Underpinning the program were some of the principles common to New Urbanism's approach to development. Namely, rather than focusing on resettling the faveladors, as had been common policy previously, the Favela-Bairro program not only aspired to introduce infrastructure and services into the area, it also set out to establish neighborhoods.[9] The practical focus of New Urbanism comes from a structuralist understanding of social life. It contends sociality is structured by the quality of the physical environment. For this reason those working within this model strive to overcome the social alienation and physical disintegration of the modern city by creating an environment defined by human scale and characterized by neighborhoods, low-density housing, green public spaces, public architecture, mixed-use buildings with more than one use (for instance commercial and residential), and a strong transportation system, all of which are stitched together to form a coherent urban fabric that aims to revitalize a sense of belonging.

Offsetting the more top-down approach the government took when drafting the policies of Favela-Bairro, Jáuregui architects used grassroots tactics during the design phase. This consisted of long walks throughout the area mapping specific social dynamics and producing different readings of these after conversations with favela leaders in what they described as a series of "mutual interrogations" between themselves and residents.[10] The question of land ownership, issues of mental health, along with microeconomic approaches and aesthetic preferences all shaped the final design. Rather than attend to the structural relationship between the physical conditions of the favelas and how this might determine cultural behaviors and sociality, the grassroots strategy of Jáuregui's team embraced the informal conditions of the region, especially the improvisational character of favela spaces and vernacular architectural styles specific to the area, drawing upon local conditions and the idiosyncratic architectural styles and materials of the makeshift shanty homes. For instance, the pathway built from the top of the favela to the basin was an appropriation of an already existing impromptu path carved out over time by people traveling up and down the hillside. The relocation housing at Fernão Cardim drew on the color, texture, and materials common to many of the houses built at the basin of the favela, as well as sharing a similar density without compromising the intimate scale already defining homes in the area.

The sustainable focus of the architecture team came from the realization that "architecture serves a social purpose . . . it cannot afford to be disliked by the community, and . . . it must be understood to be accepted, maintained, and kept functioning by the population."[11] As such, new streets were built, walkways paved, day care and community centers that also function as job training centers or communal kitchens constructed, abandoned buildings such as the Vidigal sports center renovated (the sports center later hosted the Rio soccer championships), new public space and health care facilities built, community laundries that serve the double function of a spontaneous car wash facility established, sanitation services introduced, sewer systems installed, power lines buried underground, and dirty rivers rehabilitated. Brooke Hodge reports that in Fernão Cardim the river was used to "link the city to the favela and to knit together the two sides of the favela . . . Bridges were added to maintain continuity and communication between the two sides of the community."[12] All these projects were completed with the help of local labor and with a view to creating a sense of stewardship among local residents.

The positive evaluation of the Favela-Bairro program by government is one reason it rapidly became a showcase and model for upgrading other urban slums around the world.[13] For instance, the Secretary for Rio de Janeiro Housing, Solange Amaral reported: "Evaluations of Favela-Bairro suggest that communities that have taken part in the programme have improved significantly in terms of coverage of drinking water, sewerage and garbage collection services."[14] One cannot deny that there is a lot that is commendable about the project. Although I am not interested in dismissing the Favela-Bairro project, I would like to establish where its specific strengths and weaknesses lie in an effort to ascertain how the former might be reinforced. The evaluation begins with questioning what system of knowledge accompanies the process of urban integration; how this forms a discourse that in turn makes sense of and constructs the slum dweller's body as synonymous with the informal city; and whether the model of urban integration a constitutive constraint that produces, as Butler argues in *Bodies that Matter*, the "domain of intelligible bodies . . . as well as a domain of unthinkable, abject, unlivable bodies"?[15]

Butler insists the body is always mediated by language. This is not to suggest the body only exists as language or as a passive entity that comes to life once it is signified. Rather, the body is not a natural category but a cultural one; it functions as a normative practice, producing bodies through a process of reiteration and exclusion. As norms are reiterated and unconventional bodies remain in the shadows, identity is established. Butler makes clear that identity is a construction and also harbors within itself that which it excludes, such as the shadowy, abject realities that exceed signification. Regulatory norms produce a normative materiality as they simultaneously consolidate this other aberrant bodily life.

The discourse of reiteration and exclusion produces the effects it names—the materialization of a norm—generating what it regulates and constrains. The political question this prompts is whether or not the unthinkable body can create unpredictable significations, which in turn constitute an outside discourse, albeit one that is internal to the dominant discourse framing and stabilizing the materiality of the body. More specifically, in the context of the discussion here, this invites us to consider whether or not the negative effects of slums, as signified by informality and crime, can be effectively countered by integrating these areas into the formal city. The abject position the squatter occupies is reiterated through the authority the formal city holds. Naming—the "slum," the "informal economy," and/or the "favelador"—could also be understood as establishing a boundary between normal and aberrant urban life, this being a normative distinction that persists long after the upgrading project has been completed (as evidenced in the negative perception of faveladors that UN-HABITAT noted). In addition, the follow-up reports and studies, such as the one compiled by Fabio Soares from the International Poverty Centre and Yuri Soares from the Inter-American Development Bank, also participate in this reiteration process, using the ascription of the slum dweller's body as an unsanctioned indistinct entity to bring into focus the formal city as a legible and legitimate entity.

Using census figures, Soares and Soares assessed the outcomes of Favela-Bairro by comparing and contrasting the profile of an average favelador pre–Favela-Bairro and post–Favela-Bairro with the other citizens of Rio. Prior to Favela-Bairro the decision to live in a favela was based on financial and geographic considerations. The proximity of the favela to places where the favelador works alleviated many of the costs and time associated with commuting. This offset other negative amenities, such as a lack of public services and security. Other expenses, such as contributions to public utilities (lighting, water, property taxes, and so forth) were also eased by virtue of living in a squatter settlement. Their research found that household heads whose commute to work was under one hour were 16 percent more likely to live in a favela.[16] Similarly, migrants from outside the state of Rio were 11 percent more likely to live in the favela as compared to other low-income groups.[17] They claim this profile remained largely unchanged after the Favela-Bairro program. Although real estate values in the favelas had improved, the concluding remarks of their report heeded on the side of caution, suggesting the increase in favela property values post–Favela-Bairro was part of a much larger trend occurring throughout the city as a whole.

According to Soares and Soares the common profile of a favelador is a "non-white, unemployed, migrant" who on average has shorter commute times to work than other low-income communities.[18] The report maintains this identity remained intact regardless of the Favela-Bairro upgrading program. In so doing, they represent the identity of the favelador as a natural category, one that is immune to changes in context. It is therefore unsurprising that their findings noted that crime levels had not decreased, nor had mortality rates, and overall the effect of the program was minimal because the identity of the favelador remained consistent. This view assesses and measures the slum dweller's body as part of a signifying practice, rendering that body intelligible only insofar as it exists as an inversion of the law-abiding, white, educated, national body of the Cariocas (non-favela Rio citizenry). For these reasons their assessment of the success or failure of Favela-Bairro participates in the performative production of the favelador body in a differential operation. This differential constructs the singularity of favela life as a foreign informal territory, one that is defined in opposition to the formal city. Yet the informal city also defines the formal city as its constitutive Other.

In their research on urban planning policy for poverty alleviation in the favelas, Jorge Fiori, Liz Riley, and Ronaldo Ramirez acknowledge the complex interaction occurring between culture, local identity, and social life in the favelas. They shift their focus away from a reductionist approach (defining the identity of the favelador in the way Soares and Soares do) onto an analysis of what the cause of urban poverty is. They insist the discourse surrounding urban planning and design is inherently problematic. They criticize the "uniformity of policy, language and approach" that perceives favela residents as a homogenous entity.[19] For them the policy framework of the Favela-Bairro program (used during the design phase) was not the product of participatory research involving community and nongovernmental groups. This discourse is defined by unequal relations of power that marginalize the poor. In their report they address the complex variety and combination of conditions underpinning poverty. In so doing they argue that residents and community organizations need to be involved from the outset so that policy measures intended as a vehicle for physical and social integration are the result of participatory research involving residents as well. They hit upon an important point: poverty is a discursive construct, not a natural category. This is a viewed shared by the UN-HABITAT, which in the *State of the World's Cities 2006/2007* noted: "Highly centralized systems of governance cannot benefit the urban poor if they are being run by regressive, anti-poor political leaders and inequitable policies."[20]

In this manner, the program for Favela-Bairro compiled by the Municipal Government of Rio is not neutral; in fact, it produces what it defines, namely, the state maintains a position of authority and legitimacy over and above that of the faveladors. Fiori, Riley, and Ramirez insist from the outset Favela-Bairro was already implicated in the cultural inscription of the favelador body as uniformly poor. The decision-making process took a top-down approach that involved minimal avenues for power sharing with favela community groups. And given the history of authoritarian State rule in Brazil, along with the hostilities this created between government and poor communities, they point out Favela-Bairro was implicated in power relations specific to this history from the very beginning.

The quantitative method used by Soares and Soares undermines deeper structural inequalities shaping life in the slums. Fiori, Riley, and Ramirez draw attention to the manner in which preexisting historical hostilities construct the favelador in opposition to the formal city, explaining the reason social development projects were less successful than the physical components of Favela-Bairro:

Political disputes within and between departments, limited financial resources available for social projects, the historical tendency for most departments to ignore *favelas* and the continuing reluctance of some to broaden their constituency, policy mandates that clash with the Favela Bairro approach, and a certain reluctance within the Housing Department itself to give up control over such a politically important and visible programme as Favela Bairro.[21]

The weaknesses Fiori, Riley, and Ramirez identify occur at the level of multisectoral planning, a position that resonates with UN-HABITAT, who also highlight the importance of creating decentralized forms of governance and improved intermunicipal coordination.[22]

Fiori, Riley, and Ramirez note:

If one dimension of poverty is social exclusion, then the concept of city scale addresses this dimension if signifying the city as a dynamic system through which power and decision making is devolved and equal access to opportunities and resources are ensured.[23]

As the government was establishing the design brief and program elements of Favela-Bairro, there were no avenues for power to be shared with the faveladors, nor were there mechanisms in place for them to participate in what direction or emphasis the program would have during the early stages. Although the design team's grassroots approach during design phase attempted to address this problem, if we follow Fiori, Riley, and Ramirez this did not rectify otherwise historically constituted structural imbalances of power (formal versus informal city) defining the parameters of the program prior to the design phase.

Davis notes, "Not all urban poor, to be sure, live in slums, nor are all the slum dwellers poor."[24] Similarly, Perlman reports "not all the people in favelas are poor and not all the urban poor live in favelas."[25] Both draw attention to the fact that slums are at best partially adequate to the poor they are said to represent. Depending on how the term is mobilized, the slum can represent poor identities; be used in racist discursive regimes; or erase the overlap between monopoly capitalism and racial and economic histories that together situate the formal and informal city differently from one another. Unlike the coherent, unchanging profiles Soares and Soares produced, Davis and Perlman suggest the poor do not easily fit into a homogenous economic category, geographic identity, or set of social behaviors.

After having lived in the favelas for 18 months from 1968 to 1969, at which time she interviewed 750 people in three different favelas, Perlman returned thirty years later and reinterviewed 41 percent of her original study participants, including a random sample of other members of their families. The 2,182 total interviews conducted over this time later noted vast improvements in the material conditions of everyday life in the favelas. However, even though the legal status of the favelas changed—many had gained de facto tenure, access to basic services and infrastructure— Perlman explains that the area was not successfully integrated into the broader urban fabric because of the negative perception of faveladors. This stigma can be summarized briefly using the following dichotomies: normal/indecent, honest/criminal, and neighborly/threatening.

For Perlman the real source of social exclusion was the stigma that living or having grown up in a favela carries. Hence, despite improved education this did not translate into better work opportunities for faveladors. Her coworker, Ignacio Cano clarified: "Our research shows that the favelados continue to suffer discrimination in the labor market and that unemployment has risen."[26] Perlman makes a convincing argument in favor of the stigma thesis when she cites the case of sixty-five-year-old Dona Rita from the favela Nova Brasilia, who now lives in a high-rise condominium outside of the favela. Rita owns two stores, a beach house, a car, and truck— far from being destitute—and yet while she was shopping for a new pair of eyeglasses in an exclusive downtown store in Rio, the salesgirl curtly informed her she could never possibly afford to buy the glasses that had caught her interest. Rita explained that she was being judged for dressing as a favelada—someone from the North Zone favela.[27] In Perlman's opinion the negative perception Cariocas have of faveladors seriously compromises the planning objectives of urban integration and frustrates the possibility for Favela-Bairro to have any long lasting impact.

Perlman's findings are also at odds with Soares and Soares, who explicitly state: "perceptions surrounding favelas have in fact 'improved' over time, a regularity reflected in the changes in public policy also."[28] In her book *The Myth of Marginality,* Perlman "showed that Rio's favelados were not marginal to the economical, political, social and cultural system, but tightly integrated, albeit in an asymmetrical manner."[29] This is because on the whole their participation in civic life is minimal at best. It is this asymmetric position the favelas hold within the urban fabric whereby their socioeconomic life is best understood as a singularity as compared to being an informal city. The favelas are certainly not the antithesis of the formal city; they are a reciprocal entity dynamically distributing state-centric or market-driven organizations of urban life, pushing such organizations to their limits.

In her later study Perlman notices a dramatic change in levels of violence pre– and post–Favela-Bairro, especially the expropriation of space by "drug-related gangs, vying with each other for control of the turf."[30] In particular the favelas had become even more militarized post–Favela-Bairro as drug lords "engaged in armed battles with the police (who are also complicitous in and benefiting from the drug and arms sales within the favelas)."[31]

In her study Perlman observed that after Favela-Bairro violence had escalated to the point where people were too scared to use the public spaces built for them out of fear of being caught in the crossfire between police and drug gangs. Fiori, Riley, and Ramirez locate this conflict in the historical animosity and distrust between favelados and government officials. This is a national history defined by state corruption and state-sanctioned violence, along with unequal distributions of wealth and power, not to mention unprotected tenure. These historically constituted conditions not only persisted after the program but, as Perlman's study claims, they intensified. In 2005 the prolonged and bitter warfare between police and favela drug lords prompted the deputy governor, Luiz Paulo Conde, to propose a separation barrier be built between the informal and formal city. I see this as an extension of the logic of separation underpinning the notion of urban integration because both the proposition to wall off the slums and that of integrating them into the formal body of the city rely upon the following social codes: Rio/favela, formal/informal, Cariocas/favelador. And just as Conde's proposition involves the militarization of life, so too does the notion of urban integration.

Another way of understanding the policy of eviction and resettlement during the 1960s and 70s is to see it as an extension of Brazil's militarized regime as it set out to establish itself at the center of state power. The policy to eradicate the slums was a way in which the new military regime could discipline the disruptive and independent energies of the slum dweller's body by transferring these into the managed environment of the projects. As the favelas reemerged, in large part because the favela economy strengthened, this not only exemplified outright lawlessness it also instituted a self-sufficient system of law and retribution. The social realist film *Cidade de Deus* (City of God) by Fernando Meirelle chronicles the militarized economy and life of the favelas during the 1980s, depicting their harsh, edgy, unforgiving character; streets were set to the pulse of adrenalin-filled children armed to the hilt, corrupt police, and drug dealers. The situation was one of all-out warfare. Not much has changed on that front. As *Cidade de Deus* makes clear, what motivates these kids to join the drug ring is not simply money, it also involves the benefits of being in a highly structured and paternalistic social organization, albeit one that is hierarchically defined by macho displays of power and a fierce sense of loyalty, belonging, and security (albeit conditional).

As the drug economy became law throughout the favelas, the authority of the state and the free market economy were locked out. A new power base was established, with the labor of the drug lords articulating a fundamental ideological struggle (informal versus formal socioeconomic life) through the shared medium of a militaristic master code.[32] When the military regime fell and the Favela-Bairro program set out to reorganize the physical structure of the favelas by widening streets and paving footpaths, the favelas were opened up to the formal economy of Rio. These changes in the physical structure of the slums allowed the newly formed state government to enter the area and take control. It is therefore important not to ignore the fact that the model of urban integration posed by the Favela-Bairro project was complicit in the policing of favela spaces.

Viewed in retrospect, the upgrading initiatives attempted to normalize and regulate what was once perceived to be a hostile and aggressive part of the city by reducing the gap between the favelas and Rio, targeting the rhythm and flow of life on the ground. Opening up the streets to generate smoother traffic flow was one way of alleviating the perceived threat of favela spaces. It was also an attempt at decentralizing the loci of power and resistance embedded within the raw material of favela life (the drug lords). For instance, the walls inside the favelas once carried a subversive function; often erected by drug lords who wanted not only to protect their turf from other gangs but also to impede the presence and mobility of police in the area.

Thomas Barnett stresses in his discussion of global security that nonstate actors, such as drug traffickers and terrorists, are part of a fourth-generation threat that largely relies upon the use of asymmetric tactics. The security challenge arises between what he calls a functioning core—where multinational capitalism, liberal media, and a sense of collective security are thick—and an otherwise nonintegrated gap, where globalization is thinning.[33] Barrett's solution is not one of containment; rather, the formal sector of the core has a responsibility to shrink the gap. Considered in this way, the urban integration policy of Favela-Bairro seems no different to Barnett's defense strategy.

It is therefore crucial for urban designers and planners to recognize that the negative conception of the informal economy as one in need of integration is in itself an ideological idea. Ideologically speaking, it conceals the autonomy that once came from being off the grid. In being off the grid, the favela tested the social and economic arrangements defining the profitability of the urban center and from this perspective, urban integration is an act of cultural violence; it represents an otherwise ambivalent urban condition in manageable form, one that can be absorbed by the expanding markets of capitalism.

What I am suggesting is that urban integration is conditioned by violence, and when this is traced onto areas that have a history of state militarism this condition is amplified. For this reason I am critical of the position Peruvian economist Hernando de Soto advances in *The Mystery of Capital*.[34] He contends that the way out of poverty is to formalize the informal economy so that poorer economies can be better positioned to follow the course taken by Western market economies. As long as the assets of the poor remain dead capital (for example the squatter tenement that has no real value in the formal real estate market), de Soto argues, it will be very difficult to break the cycle of poverty. Dead capital is not an asset that can be borrowed against, so the key is to integrate dead capital into the formal economy, enabling the production of wealth. His position not only takes free market and financialized capitalism as a given socioeconomic good, it also does not consider the effects of integration, which in the case of the favelas has stimulated police corruption, state-sanctioned violence, and social instability.

In addition, what happens to those living in the favelas when the energies driving their singular economies are placed in the service of the free market? As the singular economy is integrated, not only do the city and state benefit from new sources of tax revenue, but all life is placed in the service of some of the most reactive values that the free market propounds. Where once electricity and water were free in the slums, they now become commodities. For instance, after the success of the small kiosks Favela-Bairro helped establish, McDonalds saw a wonderful opportunity to expand its market base and has now become the first fast-food chain to set up shop in the favelas, in direct competition to local entrepreneurial efforts. Lastly, as Perlman described it, the favelas she studied now have a "thriving internal real estate market for rental and purchase, with prices in the most desirable favelas of the South Zone, such as the famous Rocinha, rivaling those of regular neighborhoods."[35]

The principle of formalizing the singular economies and sociality of slum dwellers relies upon a differential logic—the informal city in opposition to the formal city. From here, urban integration uses what is understood to be the normative urban condition of the formal economy to reorganize the informal sector and suspend these contradictory forces without damaging the social and economic life of the formal city. Hence, the local cultural practices and social relations specific to the slums are invariably co-opted and changed through this hegemonic model. In this light, the problem of sustainable urban design has to do with how it can first of all distance itself from this and shift the discursive parameters defining the normative body through the differential condition of the slum dweller. This is not a problem of content (introducing infrastructure, services, mobilizing credit, and so forth) but a formal problem of disjunction. It would be one that uses the boundary between the favela and Rio as a decommodifying force, celebrating the fact that this force distributes and differentiates state control and the logic of the free market economy.

Neuwirth points out:

For decades, national, state, and local governments steadfastly refused to provide services to these communities (favelas of Rio de Janeiro). And with that neglect came criminality. So most of Rio's favelas are now controlled by highly organized and extremely well-armed drug gangs. These gangs are both criminal and communitarian. They offer squatters a trade-off. In a city where assaults and violence of all sorts can be common, there is no crime in the squatter communities—as long as people look the other way when the dealers are doing their business.[36]

Neuwirth's observation seems to suggest that the drug lords are like a quasi-sustainable urban effect produced in response to very real economic and social difficulties.

The violence of the drug lords is not radical enough; it does not call into question the status of violence within the definition of the favelas as an urban condition that is held in stark contrast to that of the "formal" city. In effect the violence of the drug lords is appropriated by the state to further the state's own economic and political agendas. The police line their pockets on the back of the drug and arms trade, and newly installed infrastructure and services throughout the favelas allow the state to tax the poor and reap the financial gains of legalized electricity and water supplies. In this system patriarchy flourishes because the constant association of the favela with informality and unproductivity helps justify the domination of the favelador's body by the state and the free market. Patriarchal values such as competition, autonomy, and individualism now prevail; as they do, the free and open systems of labor and exchange indicative of the singular economy, which are necessary to the collective survival of the faveladors, are utilized for profit accumulation. In effect, urban integration views the singularity of socioeconomic life in the slums as offering new opportunities for the formal market to expand. The naming of this free and roaming life system as "informal" is the first step toward organizing the realities of favela life into a general form. As a label or name, *informal* systemizes the singular dimension of favela life by giving it a fixed reality and determinate condition.

If our current historical conditions are one of global multinational and financialized capitalism, corporate culture, skewed socioeconomic positions that cannot be disentangled from other global structural disparities, and the increasing militarization of public space, then this is also the context in which urbanism works. Therefore, if urban designers and planners are serious about unraveling or interrupting the unequal power relations that such a context supports, then their work has to somehow broach the broader question of how the discipline itself is implicated and even advances this system in order to avoid being compromised by it. When design uses the conceptual apparatus of a dominant formal city as its primary point of reference, into which it will integrate the supposedly informal city, it erases the singularity out of socioeconomic life. The utopian dimension of the favela emerges at this formal level of disjunction. The life of the favela is at once specific (arising out of the active economic and social connections with the rest of Rio and its particular historical conditions that differentiate its cultural life) and also the effect of a more universally shared historical situation (the growth of slum neighborhoods worldwide and the expanding free market economy). Indeed, it is out of the formal failure to fully "integrate," as signified by continual guerrilla warfare, where the utopian praxis of favela life roars forth. However, the proposal to build a slum wall in response to this also suggests the emergence of a total state war.[37] Reproducing the exploitative organization of life, the slum wall affects the entire population, albeit unevenly.

In response I suggest that a form of machinic urbanism needs to take hold in the planning, design, and development of slum neighborhoods. The machinic is a sociotechnological machine that refers to the pattern of nondiscursive practices organizing bodies.[38] Rather than play with the signifiers of urban organization, machinic urbanism would experiment with the material movements of life in all its variation. This materiality is immanent to the singular dimension of its social, economic, political, and cultural context as well as the ecological specificities of the area. This is neither a humanist nor organicist approach because it has no pregiven foundation or telos on which the design is based.

A machinic urbanism would focus on the productive connection of life, all those ways of life that elude control and clog the arteries of everyday life, transforming these in the process. According to this view, life is a machine, but it is a machine without identity. It comes to life by virtue of the connections it produces. The challenge for sustainable urbanism in the context of the slums is how to engage the creative lines of flight defining the social field as a whole; working to connect the micro with the macro, the public and private, and the physical conditions of the city with the flows of economic markets and political materialities (policy, ideology, history), all in an effort to effect new social, economic, political, cultural, and ecological conditions.

What I am referring to is not an everyday urbanism that is "idealistic about social equity and citizen participation, especially for disadvantaged populations" in the manner of Margaret Crawford.[39] Neither is it the post urbanist position Douglas Kelbaugh identifies with the work of architect Peter Eisenman; this being a poststructuralist approach to urban design, one that "strives to embody, accommodate, and express the contemporary forces of technology, culture and flow."[40] Nor is it simply the re-urbanist response that Barbara Littenburg and Steven Peterson are involved with as they address the "larger patterns of city, neighborhood and of public space, which jointly serve as the context and the enabler of architecture."[41] It is, however, a little of all three. It recognizes, like the re-urbanists, that there are certain immutable spatial and formal urban conditions but, unlike them, believes that these do not consist of the plan/view, horizontal/vertical, old/new, object/void dialectic. A machinic urbanist would posit that the immutable spatial and formal elements re-urbanism speaks of are indeed historical, social, material, and cultural constructions that can give rise to different outcomes at a discursive and formal level.

Machinic urbanism shares strong sympathies with the nonnormative approach of everyday urbanism, in particular the idea of attending to "not only on what is present in the banality of everyday life, but also on what is absent yet might be there," but it also aspires to radicalize the notion of everyday life to include not only human life but also ecological life, economic life, and so forth.[42] Where everyday urbanism attends to the interaction between mundane places and the overall urban fabric, machinic urbanism denies the fundamental separation between the everyday life of the city and the city as a whole in an effort to tangibly activate multiple urban scales, conditions, and forms. The two images of the city everyday urbanism produces are grounded in the perception that the two cities are separate from one another and, as such, although it engages with hidden urban meanings—freeways, the mall, or adhoc markets—it remains blind to the contribution, say, that the homeless tucked into the nooks and crannies of these everyday spaces and ways of life make to the urban fabric. Therefore, in many ways the interpretative focus of everyday urbanism perpetuates how dominant power relations distribute and define the meaning of urban spaces by simply trying to reform the separation between formal and informal urban patterns. Against the everyday urbanist model of communication that aspires to help minorities speak out, machinic urbanism prefers to revitalize the concept of everyday labor—understood from a social and ecological perspective as the flows of matter, affects, energy, and power—deepening the ecological potential of the concept in order to consider how everyday labor from the property owners to the homeless makeshift shelters of street kids, to the street vendors, and even to the flows of sewerage and water runoff, all these urban elements produce and mediate the urban landscape.

If some of the most serious threats people face when living in urban slums arise out of inappropriate land use and increasingly degraded environments, then earthquakes, landslides, floods, and drought are serious hurdles that must be addressed for development to be sustainable. One example of a form of machinic urbanism would be for the designer to engage with the slum's site infrastructure (the intensive topography of the site). For instance, Steve Luoni looks to systems at work within a given site to discover their immanent design potential.[43] Rather than restore a concrete wall staving off the hydrologic force of water on the move, Luoni prefers to revitalize the hidden path of the stream as it travels underground and turn this movement and labor into an ecological service that benefits the site and surrounding areas as a whole. He does this by allowing the stream to exist as close as possible to its natural state, incorporating the wetlands and flood planes that once kept the system of flooding in check into his design and development of parking lots, public space, and building construction. The point is that the built environment and the natural flows of the urban metabolism are not defined against one another but instead strengthen and revitalize each other.

Machinic urbanism would revive the following claim Marx made in *Capital*: "Technology reveals the active relation of man to nature, the direct process of the production of his life, and thereby it also lays bare the process of the production of the social relations of his life, and the mental conceptions that flow from those relations."[44] Another good example would be the Viva Favela portal, started in July 2001, which aspires to subvert the cultural abjection of the favela in mainstream media. It consists of favela correspondents and journalists who report on and document an alternative discourse to the one prevailing throughout popular culture, using the dominant position the media holds against itself. Viva Favela holds the media accountable for the stigma that discriminates against faveladors and renews and transforms the popular image of the favelada.[45] This is a form of machinic urbanism at work as it presents an affirmative image of life in the slums going on to generate a different discourse to that of formal versus informal urban life.

Machinic urbanism necessarily separates itself from the banal realities of everyday life, and in the spirit of post urbanism it attempts to provide a platform from which to imagine a future different from the present. Finally, machinic urbanism aspires to create conditions of agency without translating these into the superstructure (which has no autonomy), all the while maintaining a position of relative independence within urbanist discourse insofar as it is in excess of the whole notion of urban design and planning. For instance, the Favela Roncinha Tour could be described as a machinic urbanism. It is the brainchild of Rejane Reis and attempts to develop a sense of agency among the youth of the slums using the concept and practice of "touring" as a discursive medium. The tour skirts the masculinist paradigm defining the patriarchal structure of the drug cartels and police corruption, opening up the streets to a different social life. For just $25, locally trained guides give tourists an insider's look into one of Rio's largest favelas. While touring the area, people also support the local economy when they take the opportunity to purchase locally made handcrafts.

This tourism project sets out to change people's perceptions of Rio's slums and create job opportunities for youngsters between fourteen and twenty-five years old. More important, the project targets disadvantaged and vulnerable youth (many of whom would end up working for the drug cartels) by improving their self-esteem and building a sense of pride and stewardship for their neighborhood. The young guides develop language skills (English and Spanish) and learn about other places in the region that are of interest to tourists (Corcovado, Sugar Loaf, and Tijuca Forest). Reis reports: "When they arrive they are shy and can barely speak. Throughout my Tourism Workshop they even learn how to smile!"[46]

The Favela Roncinha Tour shares a great deal in common with microeconomic schemes that begin with the following premise: for change to be effective in slums, cultural change is necessary. The program is successful in a nonideological and nonsymbolic way because it affirms the abject singular layer of the urban fabric (the most defenseless members of society) without reifying its social life. What is more, it draws attention to the gap between the exploitation of childhood labor in the favelas and the patriarchal structures (the police and drug lords who profit off that labor) that define such exploitation, all the while wrenching childhood labor away from these oppressive circumstances. To the extent that the program aims to develop a child's sense of confidence in their abilities, it reinvents the concept of profit and the bottom line, shifting the definition away from increasing profit margins onto one of emancipation and making an investment in the future (children).

Até Quando (When will it end?), a multimedia program set up by the nongovernmental organization Observatório de Favelas, also sets out to make the favela and its culturally specific way of life visible as a social subject and to empower favela youth. The project is part of the much broader grassroots initiative Popular School of Critical Approach to Communications, which offers underprivileged youth courses in photography, community radio, video, and print media.[47] Armed with a camera and not a semiautomatic, young people involved in this program document the world around them. As they repeat the sociopolitical forces of their neighborhood in visual form they bring into existence a different image of the area. Also, by tapping into more productive forms of power, as opposed to the macho displays of power driving the drug wars, in this program children are considered an important resource not just for the future but for producing change in the present. The perspective children bring on favela life in itself renews our understanding of what constitutes an urban resource. Alive to the favela's concrete vitality, they present a collective vision filled to the brim with the variety, color, noise, and texture of the favelas. This perspective changes not only how they understand their own lives but also works to modify the perception others have of favela life.

The power of sustainability culture in this context comes from the way it agitates the dualistic framework underpinning popular representations of slums and slum dwellers. Accordingly, the opposition between formal/ informal, civilized/barbaric, city/slum, legal/criminal, organized/chaotic, state/anarchy, and planned/unplanned is ultimately exposed as being ideological, not real. Indeed, design initiatives such as those by Jáuregui architects, regardless of their good intentions, fail at the level of form because they are premised upon such oppositions, and these can only ever be ideological. In so doing, the focus on rendering the favela more manageable strips the political impetus right out of the hands of the very people it sets out to empower. Worse still, the favela is turned into a space of risk in which the task of the designer is to effectively take control of such risk in the present so as to ward off the perceived threat of what the future may hold. Putting such an ideological opposition to work runs the risk of reorganizing the favela space so that it is transformed from a semi-autonomous entity into a bureaucratized space. There is a fundamental difference between the ideological construction of urban integration that only ends up reinforcing state and masculine displays of power (the war between drug lords and police), and independently constituted urban alternatives that attempt to shift the power base into more collective structures such as those offered by Reis and Até Quando.

One way for a community to become more robust is to build upon and expand common experience, which necessarily means history must be embraced as both the foundation and absent cause of any sustainable design initiative. To do this, urbanism somehow must engage with the normative criteria implied within the vocabulary of urbanist design and planning and with the manner in which this brings a subject into being, concomitantly accommodating the unpredictable dimension of subject constitution as part of its practice. This claim is actually about being always "already lost to or always already expropriated by a past discourse" that we have no control over, and the future of a discourse that we cannot control either.[48] As Butler describes, this is a "certain principle of humility and a certain principle of historicity that exposes the limits" of autonomy but is also the condition of autonomy.[49] The temptation to pigeonhole slums as informal urban spaces, a classification that also inserts the slum into a normative idea concerning what is a proper mode of urban production, is dislodged. Allowing the historical experience of different and even contradictory modes of urban production to exist together may just allow those contradictions to move to the forefront of social, economic, and political life. The task of machinic urbanism, thus understood, is to transcend the oppositional logic of informal and formal cities, macro and micro scales, public and private domains, human and nonhuman life—simultaneously rousing the material realities that energize these oppositions, putting them to work toward a logic of collectivity.

9 Poverty

On average, someone living in a developed nation consumes twice as much grain, twice as much fish, three times as much meat, nine times as much paper, and eleven times as much gasoline as someone living in a developing nation.[1]

Consumers in high-income countries—about 16 percent of the world's population—accounted for 80 percent of the money spent on private consumption in 1997—$14.5 trillion of the $18 trillion total. By contrast, purchases by consumers in low-income nations—the poorest 35 percent of the world's population—represented less than 2 percent of all private consumption. The money spent on private consumption worldwide (all goods and services consumed by individuals except real estate) nearly tripled between 1980 and 1997.[2]

From 1996 to 2006 the earth's surface temperature was reportedly the hottest on record (since 1850, when the recordings began), with average Arctic temperatures increasing twice as fast as the overall global average. Tim Flannery warns that if we continue with business as usual, over the course of this century there will be an average increase of 5 degrees Fahrenheit in Earth's climate (give or take 3 degrees).[3] This situation has resulted in thermal expansion and melting ice and snow, gradually causing sea levels to rise. Flannery points out: "While the scale of change is less than that seen at the end of the last glacial maximum, the fastest warming recorded back then was a mere 2 degrees Fahrenheit per thousand years. Today we face a rate of change thirty times faster—and because living things need time to adjust, speed is every bit as important as scale when it comes to climate change."[4] More important, if human activities continue to release significant amounts of heat-trapping greenhouse gasses into the atmosphere, this phenomenon will not only persist but compound. To state it bluntly, under these circumstances the future looks bleak. What eventually will accompany the warmer weather trends are vector-borne diseases, retreating ice caps, and species extinction. The rising sea levels will erode coastal areas and small island nations will quite simply vanish. Extreme weather events, such as flooding, drought, fire, landslides, and hurricanes will increase in intensity or frequency, seriously compromising agriculture and the security of food supplies. And if the world's poor are currently "trapped by disease, physical isolation, climate stress, environmental degradation, and by extreme poverty," global climate change will certainly hit them the hardest.[5]

To set the record straight, the Earth's temperature is, unquestionably, warming. Human activities—deforestation and the burning of fossil fuels to name a few—are, also unequivocally, responsible for the mounting concentration of greenhouse gasses in the atmosphere. The question that remains is what we are doing about it.

Monbiot boldly declares: "It does not matter whether we burn fossil fuels with malice or with love. As far as the atmosphere is concerned, it is not concerned."[6] The solution he suggests is to introduce a total cap on national carbon emissions, a system that can distribute these emissions, and a 90 percent cut in the emissions of the economies of wealthy nations. He argues that we no longer can afford to surrender the future to an "aesthetic fallacy" by choosing solutions that appeal to our tastes, "rather than those which work best."[7] Monbiot is in favor of extreme pragmatism, maintaining we cannot affirm life without concomitantly destroying those cultural beliefs that negate it. For instance, the affluent are merely being tokenistic when they buy ecofriendly products, eat organic, and drive a hybrid car while they also continue to consume as much as their income allows. In effect, Monbiot is demanding an ethical choice between life-negating nihilism and life-affirming creative energies that have the potential to change the world.

Since the 1990s, popular culture has become more committed to pragmatically addressing a myriad of problems directly and indirectly associated with contributing to greenhouse gas emissions. Because of their practical focus, the design disciplines—urbanism, architecture, planning, product design, and graphic design—are especially well positioned to make a positive contribution to the problems associated with climate change. Those involved in "green design," of which ecoefficient and ecoeffective design are subsets, use their skills and knowledge to produce objects or services with minimal negative impact on the environment. Using design as a tool of environmental awareness and protection, green design aspires to reduce the amount of nonrenewable materials used in the design process, as well as to use local materials and skills to decrease the ecological footprint of the final product or service. The service-oriented focus of green design strives to reinvent modes of consumption that otherwise would promote product ownership.[8] Furthermore, green design takes a different approach to the whole idea of consumer culture. That is, instead of producing objects that end up at the local landfill, the life of the final product is treated cyclically.

Fully aware of the difficulties that arise from treating patterns of consumption as a blank slate, in *Cradle-to-Cradle* William McDonough and Michael Braungart resuscitate an idea Walter Stahel advanced during the 70s, suggesting we reimagine modes of production by replacing the cradle-to-grave model with a cradle-to-cradle system:

products that, when their useful life is over, do not become useless waste but can be tossed onto the ground to decompose and become food for plants and animals and nutrients for soil; or, alternatively, that they can return to industrial cycles to supply high-quality raw materials for new products.[9]

For these reasons, the old linear model (cradle to grave) is replaced with one that uses waste-free production techniques, environmentally sound materials for the creation of objects and artifacts, which can in turn be up-cycled. In other words, ecologically intelligent design assesses the environmental impact of a product or service throughout its life cycle.

Contesting the worth of the ecoefficient model common to much of green design, McDonough and Braungardt boldly proclaim that the goal of "zero waste, zero emissions, zero 'ecological footprint'" is a "depressing vision of our species' role in the world."[10] They contend that human beings are not necessarily inspired by models of being less bad or more efficient, and that efficiency in itself is not necessarily a good thing—for example, a gas chamber kills people more efficiently than simply hanging them one by one. All in all, they insist "eco-efficiency only works to make the old, destructive system a bit less so."[11] Their biggest objection to the goal of zero waste is that it implies we accept things the way they are, implying we should try to do the best we can with an already inefficient and destructive way of life. They argue ecoefficiency may be laudable but that this model marks a failure of the imagination by focusing on what *not* to do instead of optimistically moving toward what *to* do. They advise moving away from the model of ecoefficiency—doing more with less—and instead applying ecological cycles found in nature to industry. Rather than striking a balance between ecology, equity, and economy by limiting consumption, they suggest increasing production in a way that is ecoeffective.

An important distinction must be drawn between what is commonly referred to as "green design" and "sustainable design." Green design uses green technologies (technologies that do not pollute or deplete resources and that are energy efficient) and environmentally sound materials (assessing first the environmental impact of a raw material and its ecological footprint, along with the emission rate of the material during production phase); the more expanded field of sustainable design grapples not only with environmental issues, but social and economic ones as well. To be exact, McDonough and Braungart favor a regenerative approach, one that promotes the capitalist model of production and consumption by drawing upon the ecoeffectiveness of ecological systems, combining social justice issues (equity), with the production of affordable, safe, intelligent commodities (economy) all the while using the ecological equation of waste = food (ecology). The production, manufacture, distribution, and disposal of the product or service are considered a resource and benefit to the community, and the overall process is seen to promote environmental and economic health. This is unquestionably an innovative way to conceive the interdependence of ecological systems and human activity. However, although they acknowledge sustainability is *meaningless* if the ecosystems and biodiversity needed to maintain the quality of life of future generations are not supported and cared for, they do not fully address the fact that sustainability is also *useless* if we reduce or restrict the dynamic and creative energy of life to profit-maximizing economic principles.

McDonough and Braungart might appreciate that sustainability culture is an ethical practice, yet they continue to preserve an economic rubric responsible for some of the most unethical distributions of power, which have resulted in only a few benefiting, disproportionately, from global capitalism. For this reason, ecoeffectiveness participates in a form of illusionism, to the extent that it continues to be spellbound by a profit-driven approach to sustainable theory and practice. As such, although involved in green design I am less convinced that McDonough and Braungart exemplify the strongest strains of sustainability culture. From a sustainability standpoint, what McDonough and Braungart propose is untenable because they bracket the onto-aesthetic aspect of sustainability culture that discovers its coordinates in a mechanism that works hard to liberate material forces (content) to invent new signs that express a different way of life. This would be a life wherein economics is no longer restricted to the content of multinational corporate capitalism and instead expresses the vital creativity of life.[12] No doubt growth is an important factor in the alleviation of poverty, and as Muhammad Yunas makes clear with his work as part of the Grameen Bank—an organization that provides for poor people without collateral (primarily women) small loans with which to launch or expand a small business—the free market can address problems such as global poverty and environmental degradation, "but not if it must cater solely and relentlessly to the financial goals of its richest shareholders."[13] As Yunus says, the key to success comes from generating a new concept of "social business," whereby the goal is not so much the maximization of profit but the creation of social benefits, services, or goods. This is what sustainability culture ultimately aspires to: creating social benefits and environmentally sustainable ways of life.

Sustainability culture necessarily releases the dynamic materiality of life so that environmental stewardship and social justice no longer carry a representational function (an adjective describing a particular cultural phenomenon such as green design). Instead, sustainability culture works to invent social modes of sustenance through collaboration, cooperation, the sharing of ideas, and consultation, all in an effort to support, encourage, and attract an investment in life with a view to reaping environmental and social benefits both for today and for future generations. The claim being upheld here is that as long as sustainable design or culture functions as a signifier of traditional capitalist values (maximizing profit, equity growth, future dividends), the creative and multidimensional materiality of life is subordinated to a particular regime of meaning—a nonsocial understanding of the free market. The effect of this is that the most vulnerable—the poor—are simply left behind. Yunus succinctly explains:

Poverty is a multi-dimensional phenomenon. It is about people's lives and their livelihoods. To free people from poverty, all aspects of their lives need to be addressed, from the personal level to the global level, and from the economic dimension to the political, social, technological, and psychological dimensions. These are not separate and disconnected elements but closely intertwined.[14]

In *The Climate of Poverty: Facts, Fears, and Hopes*, which Christian Aid released in May 2006, the many ways in which the poor have dispropor- tionately been affected by climate change is clearly demonstrated.[15] The problem is so acute that the UN Millennium Development Goals (MDGs) to halve poverty and stop the spread of HIV/AIDS by 2015 more than likely will not be met.[16] One of the most alarming statistics Christian Aid cites is that 185 million Sub-Saharan Africans are estimated to die by 2100 from diseases directly related to climate change.[17] In addition, 94 percent of disasters and 97 percent of natural-disaster-related deaths occur in the developing world.[18] Arguing that poverty and climate change are inextri- cably linked, the report explains that climate change is set to exacerbate famine, drought, floods, and war for the world's poor if additional aid is not given to help empower these communities.

Throwing our hands up in despair will not help anyone; according to Christian Aid, as long as those who have the power to make changes do so immediately (India, China, Brazil, and the countries of the G8) there cer- tainly is reason for optimism. The argument is simple: since the wealthy countries are largely responsible for having created the current situation, they are the ones who need to fix the problem by cutting their CO_2 emis- sions. Like Monbiot, the report argues for a carbon budget aimed toward achieving a two-thirds reduction in greenhouse gas emissions (from 1990 levels) by 2050. Further, were clean technologies introduced and renew- able energy sources (solar, wind, geothermal, biomass, hydropower energy) used in poor communities, these changes would have the potential to radi- cally transform the dynamic and cycle of poverty, helping people get a foothold on the ladder out of poverty, as Jeffrey Sachs might say. Energy for refrigeration would enable vaccines to be safely stored; light in schools and small businesses would extend operating hours into the evening, pro- viding opportunities to empower women—these are just a few examples of sustainability culture in the service of the world's poor that could have an enormously positive impact on quality of life.

When sustainability culture creates mechanisms that affirm the vitality and dynamic materialism of life, it also unmasks the illusionism to which a profit-driven interpretation of life panders. For instance, people in the developing world—mainly women and girls—spend approximately 40 bil- lion hours annually collecting water that is often contaminated, and as a result of water-borne diseases a total of 443 million school days are lost annually.[19] More often than not, girls spend their time collecting water instead of attending school. The PlayPump water system, designed by Trevor Field, addresses gender inequality, the problems of safe drinking water in developing countries, education, health, AIDS awareness, and employment all in one. The pump is a merry-go-round that is usually installed near a school or at the center of a community, and it puts the energy of child's play to work to extract clean drinking water from deep underground. The water is stored in a large overhead tank. The tank provides up to 2,500 people with fresh drinking water. Manufactured in South Africa, the pump is installed by local crews who receive training and a ten-year job guarantee to maintain the pump. Maintenance is paid for by billboard advertisements (in Motshegofadiwa it was an AIDS-awareness campaign) posted on the water tank. The vital materiality that the Play-Pump taps into is that of play and work. The pump might best be described as an autoconstituting independence (empowering girls, women, the sick, and unemployed) that is produced in dependence with current limits and conditions (arid land, poverty, sickness, and limited resources).

The claim that sustainability culture is a process of autonomous constitution is merely to suggest that it entails destroying conventional ways of being (reactive values) by creating new ways of putting habit to work for different outcomes (in the way that the PlayPump puts child's play to work to produce clean, cold drinking water, and where surplus value is not restricted to monetary profit but expanded to include the health and well-being of a community). The proposition is: not everything in life has to be exploited for human interest and commercial gain. As the example of the PlayPump demonstrates, as energy persists it affects other bodies and only then, through this affective investment of energy, does a collective action come about. Similarly, the grow-your-own clinic submitted by Kyoto University to the Architecture for Humanity Africa design competition achieves this. The idea behind their Kenaf Field Clinic was for communities afflicted with HIV/AIDS to plant kenaf seeds, then when those grew to create a makeshift tensile structure to be placed on top of the matured plants, producing an open area underneath in which a mobile clinic could be housed. When medics are finished treating patients, the kenaf can be cut down and eaten. The project responds to medical and nutrition needs together. The clinic project uses aesthetics in the service of collective struggle—poverty, disease, and hunger—in an effort to initiate real opportunities for change.

In effect, the aesthetic problem these two design projects bring into relief is one of how to invent new values from within the limits of present circumstances. Put succinctly, without changing the idea that life is profit driven, life will never be snatched from the jaws of reactive values—and herein lies the limitation of McDonough and Braungart. The point is not that the practitioners of sustainability culture work pro bono, or even refuse to make a living from their work, rather that sustainability not be put in the service of free market economics or what McDonough describes as "profitable sustainable product development," whereby profit is understood in terms of a financial bottom line.[20] Their position, although certainly hopeful, may only be so for the affluent.

Consider the January 2005 announcement by Nicholas Negroponte, from the MIT Media Lab, at the World Economic Forum in Davos, Switzerland. He presented a plan to help eliminate poverty through education, specifically by providing laptops to 150 million disadvantaged schoolchildren. His goal was to manufacture the sturdy wireless laptop and sell them for only US$100. The focus of the nonprofit One Laptop Per Child program (OLPC) is entirely on alleviating poverty through education by providing "children around the world with new opportunities to explore, experiment and express themselves."[21] That is until Microsoft and Intel began to compete with the OLPC laptop. As Negroponte was struggling to lower the cost of his laptop from $188 by negotiating hard with leaders around the world to commit to purchasing the product (which would also lower its cost), Intel designed a $230 version—*Classmate*—for which Microsoft would provide the software for only $3 (Windows, Microsoft Office, and educational programs).

Some might not care if Microsoft jumps on the bandwagon, as long as poor children get their laptops. The difference, of course, is that to Microsoft and Intel, schoolchildren in poor countries are merely a way to expand their market base and short-circuit possible competitors, such as Advanced Micro Devices, Inc., who supply the chips for the OLPC, from gaining a competitive edge against them in the developing world. After having established a series of informal agreements with leaders around the world to purchase the nonprofit OLPC laptop, some countries (such as Libya, which had agreed to buy 1.2 million) began opting for the Intel/Microsoft combo instead.[22] As orders for the OLPC laptop decreased, costs increased. The fear was that as Intel/Microsoft stomped out the competition, they would be able to later hike up the cost of their product. All this would do is leave the poor children of the world high and dry once more.

Negroponte's nonprofit approach was one that ushered in a new way of perceiving children, the poor, and science and technology. Children were seen as a natural resource for the future, not a market to be exploited. The position Negroponte took started from the premise that poverty was unacceptable. From here, the connection between a lack of education and opportunity in adult life was regarded as unsustainable, and the way to promote a self-sustaining individual or community was to apply design in the service of promoting sustenance again. It was most definitely not a means for generating surplus value. The most sustainable response to this situation of big business profiting off of or stealing nonprofit design ideas is to develop a copyright licensing system that protects the donation of design services but a license that is flexible enough so that it does not impede the "wider distribution of innovative ideas,"[23] something that Architecture for Humanity is collaborating with Creative Commons to produce.

Negroponte's nonprofit OLPC highlights that marriages between the sustainability movement and big business need to be treated with caution. My concern is that such a marriage may empower the already powerful without consideration for improving the lot of those countless others whose access to power is severely restricted or compromised. Worse still, it may propound the syndrome of poverty that any reputable advocate of sustainability works so hard to overcome. Under these circumstances, to ask if our goal is to starve ourselves, or to "deprive ourselves of our own culture, our own industries, our own presence on the planet" as we aim for zero, as McDonough and Braungart do, may only perpetuate the disproportionate accumulation of wealth in the countries of the North.[24] In failing to move beyond the capitalist system of production, consumption and reproduction, they treat the language of culture and commodification as neutral. But commodity culture is neither value-free, nor is it free of ethical implications. The inability to recognize the very real difference between consumption to meet basic needs, commercial consumption, and excessive consumption only sustains the power relations that exacerbate the overconsumption of natural resources and the unequal access to and distribution of these. One discourse (commodity culture) fails to translate or express the truth of another (power and oppression).

In addition, the ecoeffective suggestion that design mimic ecosystems as the basis for generating new modes of production is also one that supposes the natural world only has commercial significance and application. Vandana Shiva has provided a heated argument in response to this practice describing how multinational corporations profit from indigenous knowledge and ecological systems when they transform what is otherwise public and traditional knowledge into private property. The patent on life, as she calls it, amounts to nothing less than blatant thievery. In the opening pages to her book, *Biopiracy*, she declares: "Patents on life enclose the creativity inherent to living systems that reproduce and multiply in self-organized freedom. They enclose the interior spaces of the bodies of women, plants, and animals."[25] The impact of intellectual property rights in this context stifles the "creativity intrinsic to life-forms and the social production of knowledge."[26]

There are many questions that ecoeffective design solutions leave unanswered. For example, who will own the copyright to the new sustainable technologies? How will technology address the perennial violence many face as a result of natural disasters, poverty, or what the DoD describes as *irregular warfare* (the use of unconventional war strategies on unconventional enemies that can involve both state and intrastate actors such as rebels, partisans, and so on)? These are circumstances that motivate people to become involved with the drug and arms trade, or the trade in illegally harvested wood. What happens to the people around the world who do not have access to clean technologies or even the skills to produce or use environmentally friendly solutions? Who or what will deter designers from converting the natural capital (which they describe as ecoeffective) of ecoregions and indigenous populations into intellectual capital? More important, will they be prompted to share the profits from converted indigenous capital with the indigenous communities who have used it free of charge for generations?[27] The ecoeffective model fails to acknowledge that the world is not simply out there for us to encounter or appropriate without mediation. The world does not just exist; it is produced, and culture is an integral part of this process.

The aesthetic question this prompts is how to live a life that affirms the creative pulse of life, marking the inauguration of the artist-citizen, a person who is pragmatically involved with unleashing the expressive potential of life by looking to existing needs and from there setting out to create solutions. This is exactly how the nonprofit organization Architecture for Humanity works. At its most sustainable, culture has the creative power to raise awareness and the ontological potential to empower people and generate a sense of confidence. Architecture for Humanity operates by giving communities tools, knowledge, skills, and spaces that enable those communities to help themselves and which also promote the development of social networks. Instead of imposing a design solution onto the community, the volunteers of Architecture for Humanity cooperate and collaborate with communities to address local needs as defined by the community itself.

For instance, after Hurricane Katrina hit the Gulf Coast and Biloxi, Mississippi, Architecture for Humanity worked on the ground to provide design assistance to communities who had lost their homes in addition to directly helping others rebuild. Meanwhile, in the Bay View Model Block Project, the Architecture for Humanity San Francisco chapter is working to reconfigure the streetscapes of a San Francisco residential neighborhood crippled by gang violence, high rates of hit-and-run accidents, pollution, flooding, and overall urban blight. Working with only one block—the 1700 Block of Newcomb Avenue—this project aims to help residents reclaim the area by slowing traffic, creating public gathering areas, and reducing storm-water runoff. Collaborating with the residents in a series of community design workshops and working with the San Francisco Redevelopment Agency and Department of Public Works, Architecture for Humanity will turn the area "into a model for a sustainable community—socially and culturally as well as environmentally and economically."[28]

Architecture for Humanity realizes the following imperative: take an absolutely pragmatic leap into the abyss of unimaginableness to extract from it something that expresses the vitality of life and which then can be used to construct a sustainable way of life in the present and for the benefit of life in the future. Their work is aesthetic insofar as it involves connecting communities to the creative life of matter and social energy. It is also political because they engage, displace, and reorganize the very conditions that constitute the sustainability of design, pushing "architecture" into the background so that the vitality of social life can step up into the foreground. Finally, it is ethical because it does not negate the vitality of life; it affirms and is strengthened by life's vital and material dimensions. Architecture for Humanity constructs an expressive horizon through the erasure of what is immediately given (urban blight, poverty). Life is oriented creatively precisely because culture (creative ideas in practice) is involved with the construction of new realities.

On the other hand, the position of ecoeffective design is constituted by the very social norms and conditions that make up the logic of late capitalism. The premise is that limits to personal freedom, the market and wealth accumulation are inherently bad. Yet, as the work of Architecture for Humanity demonstrates, limits do not necessarily narrow horizons in an unhelpful way, as the ecoeffective approach to sustainability might assume. When working with the limits of what is practicable so as to arouse a sense of hope in the future and bring about tangible positive change in people's lives, limits become a source of empowerment. The onto-aesthetic potential of sustainability culture arises out of not only developing a sense of respect for limits, but also in the oppositional position such a proposition creates within the dominant system of free-market capitalism. Sustainability is first and foremost a mode of sociality—creating connections using the energies and forces that constitute the social field productively—that works in opposition to the commodification of life on Earth.

In 2002 BBC News reported, "The US contains 4 percent of the world's population but produces about 25 percent of all carbon dioxide emissions. By comparison, Britain emits 3 percent—about the same as India which has 15 times as many people."[29] U.S. greenhouse gas emissions are the heart and soul of the country's economic engine. In this regard, given the trenchancy of free-market economics it is unsurprising that, in the United States at least, McDonough has rapidly become the fashionable mentor for sustainability culture à la capitalist style. Hailed by *Time* magazine in 1999 as the "Hero of the Planet" for his utopianism that "in demonstrable and practical ways—is changing the design of the world," McDonough has been the recipient of the Presidential Award for Sustainable Development (1996), the National Design Award (2004), and the Presidential Green Chemistry Challenge Award (2003).[30]

Yet, there seems to be a real disconnect between the increasing popularity in the United States to consume green products and the need to make very real changes in a way of life defined by excessive consumption and evaluated according to the accumulation of wealth and property. Although an approach to products or services that is service- and cyclically oriented is a step in the right direction, ecoeffectiveness does not exemplify the strongest strains of sustainability culture. Firmly committed to capitalist modes of production and consumption, the ecoeffective approach of McDonough and Braungart does not attempt to transform the unequal power relations such modes of production reinforce. For this reason, I firmly believe that on its own the commodity form of ecoeffectiveness is not dynamic enough to be considered truly sustainable. Yes, global warming will impact all life on earth, but not equally—the world's poorest populations are the most vulnerable to the negative effects of climate change. Bluntly stated: in order to be sustainable, sustainability culture must be committed from the outset to empower those most at risk by climate change—the so-called other 90 percent—at both local and global levels.

Take the extreme poor living in developing countries across the world and struggling to meet their basic needs on a daily basis. According to Sachs, in 2001 the absolute poor made up approximately 1.1 billion of the world's population.[31] As Sachs outlines, it is possible to end global poverty by 2025, but it would mean that national resources could not be used for waging war, political infighting, or squandering through corruption; moreover, affluent nations would have to help more. "If West Africa has a population of some 250 million people," the $4.4 million that USAID provided over three years "would be *less than a penny per person per year*"; putting this into perspective, Sachs explains that is "enough perhaps to buy a Dixie cup, but probably not enough to fill it with water!"[32] As the practice of Architecture for Humanity suggests, for efforts to be truly sustainable they must take hold of the problem in its totality and assess where the failures lie and what the needs of a community or situation are.

Indeed, the practical value of sustainability culture comes from how it formally contests the limits of what is believed *to be* practical and realistic, and what *could be made* practical and realistic. The most challenging reality is as Arundhati Roy so poetically writes:

To love. To be loved. To never forget your own insignificance. To never get used to the unspeakable violence and the vulgar disparity of life around you. To seek joy in the saddest places. To pursue beauty to its lair. To never simplify what is complicated or complicate what is simple. To respect strength, never power. Above all, to watch. To try and understand. To never look away. And never, never, to forget.[33]

This happens by confronting honestly the difference between being virtuous and being willing to do something. The difference is not just ethical; it is also absolutely and necessarily political. After all, it is pointless to claim we support ideas of sustainability and somehow still want to retain our attachment to the accumulation of wealth and private property, all the while turning a blind eye to the unspeakable violence that surrounds us.

And if this is too complicated to digest, then a warning voiced by Sachs may help put things into perspective: "the question isn't whether the rich can afford to help the poor, but whether they can afford not to."[34] Regardless of how much Monbiot might demand we cut back on emissions and introduce a system of carbon rationing, if we continue to look away from the problem of global poverty, hunger, and disease, changes at the national or even international level to initiate environmentally sensitive policies will never be sustainable. No amount of ecoeffective design infiltrating the culture of the corporate world, as McDonough advocates, will stop the widespread phenomenon of environmental degradation and global unrest if the situation of poverty persists.

To be clear, for the future to be different from the present and past, action is absolutely critical, but so too is tempered and well-informed critique that takes the time to question the status quo and the priorities of free market economics that have disproportionately disadvantaged the world's poor. Now more than ever, as the negative effects of climate change take hold, culture needs to persevere with exercising its extraordinary ability to facilitate opportunities through which to reimagine life by speaking directly to and of the struggle for collective life in practical and inspirational terms. The key is for culture to inspire the present to cooperate with future generations by daring the present to pragmatically mobilize the limits of what is currently feasible.

Conclusion

Watching the news in December 2007 I was moved and dismayed as I witnessed a mass of green fingers fill the screen. The image was potent and precise. It condensed a variety of meanings—growth, life, environmentalism, and solidarity—in one simple gesture. These were the index fingers of environmental activists in support of the Climate Change conference held in Nusa Dua, Bali. A wry smile crept across my face as Arnold Schwarzenegger's appearance on the cover of *Newsweek* in April 2007 spinning the Earth on his finger also came to mind. But unlike the protesters in Bali, the rebranding strategy of Schwarzenegger as the new "green" governor took a very different tack. Spoken in the true words of a former Terminator, Schwarzenegger's slogan on *Newsweek* read: "Save the Planet or Else." Disturbed by the associations my memory was conjuring up I began to wonder about the power of culture. Were the images of Schwarzenegger and the Bali activists different? And if so, how? The multitude of green fingers threw me off balance. Schwarzenegger's muscle-flexing ultimatum simply left me feeling cynical. The difference was one of affect.

Holding these two images—Bali protests and Schwarzenegger—in diametrical opposition has everything to do with remaining in fidelity to the political function of sustainability culture. It is intended to distinguish between culture functioning as a point of disequilibrium and insurgency, and the more mediated form of culture that functions as a point of control and order. The former is used to enhance life, the latter to limit it. Mulling further over the distinction I realized that the power of sustainability culture is above all *potentia*. Hardt and Negri assert *potentia* is the cornerstone of political power; the onto-aesthetic force of sustainability culture is expressed through collective struggle. It is the will to collective good that took place in Bali. It is the energy of defiance presenting itself in the popular media, such as the image in Australia's *The Age* on February 4, 2007, of two polar bears stranded atop a chunk of ice surrounded by the open seas on the brink of drowning unless a miracle occurred.[1] It stirs forth in the affective combinations marking the Greenpeace campaign photographed by Spencer Tunick in August 2007, where 600 nude bodies banded together at the melting Aletsch glacier in Switzerland.

Less clear, though, is how the unruly *potentia* of sustainability culture can directly inform politics without mediation, because when mediated it risks slipping into the recesses of *potestas*, turning into a mere expression of state and corporate political power. Yet, if it remains totally unmediated does sustainability culture adopt the very oppositional political logic—core and periphery, totalitarianism and democracy, war and peace—that Hardt and Negri claim is now redundant? Do we need to choose between mediation and contradiction, or can we conceptualize the *potentia* of sustainability culture taking a transverse path through both? This would mean starting out in the middle of the two, proceeding to invert the negation and encounter a difference in itself, and then—more importantly—to tackle what this difference can do. The difference I am thinking of is that between science and art, not as distinct disciplines or schools of thought, but a difference as *potentia*—a creative force, if you will.

In 2006 the Aletsch glacier retreated 377 feet. If global climate change continues at the current rate, by 2080 the majority of Switzerland's glaciers will be gone. Polar bears are on the brink of starvation and their populations are rapidly declining, and it is little wonder given the state of the collapsing Arctic ecosystem. But how much do the statistics scientists throw at the public motivate people to make choices, compromises, or even slight changes to reduce their carbon emissions? Obviously, given the current state of the situation, they are not accomplishing much. If the predictions are not to be scrambled and depoliticized of all affect in a mire of maneuvers ranging from the corporate to the militaristic, the scientific calculations urgently need to be sensed and realized through collective experience. The statistics of climate change remain abstract unless transformed in a manner that engages the popular imaginary on an affective and emotional level. The suggestion is for the power of science to connect with the power of art. The power of art lies in the manner in which it makes the invisible force of death visible through the sensible world of affect, sensation, and emotion.

In keeping with this idea, Tunick's images of bodies strewn across rock where glaciers once lay connect the visibility of a changing climate to the invisible forces of immanent disappearance. Connecting the qualities of skin and stone, hardness and softness, cold and heat, all of which he framed with the invisible threat of impending extinction, Tunick brought together two fragile landscapes (human bodies and the glacier), vehemently and eloquently visualizing existential vulnerability to maximum effect. Poised together at the edge of extinction the human being, like the glacier, screams out at death. The invisible force of the unforeseeable future kicks in the door so that the present cannot remain immune to the future glaring down at it. Seizing upon the insurmountable and extraordinary power of futurity, Tunick confronts the present generation with its own arrogant indifference.

Similarly, Jean-Claude Didier's living installation—*Trapped Inside* (2006)—on the United Nations Environment Program grounds in Gigiri, Africa, also contracted the potentiality of death that climate changes poses into one simple, lucid image—the endangered indigenous African Greenheart tree (*Warburgia ugandensis*) on life support in an atmosphere-controlled structure. Who would have thought we ever would need to keep a tree alive in an intensive care unit? Both Tunick and Didier use the common language of art to sustain the provocative experience that arises when art encounters science, and the language of science is mediated in opposition to the epicenter of global power: U.S.–style capitalism and militarism.

Many of the case studies in this book demonstrate that the network of military and multinational corporate power extending across the globe is not supranational. Although globalization is diffused through the spread and flow of goods, capital and people, clots emerge. At the beginning of the twenty-first century, one of the the biggest clots for peace and for achieving an international agreement to cut carbon dioxide emissions has coagulated around U.S. hubris. As Patrick Tyler responded in the *New York Times* to the 6–10 million people who took to the streets in protest against the United States–led invasion of Iraq on February 15, 2003, the demonstrations remind us "that there still may be two superpowers on the planet: United States and world public opinion."[2] A case in point would be the smug demands made by U.S. representative Paula Dobriansky at the Bali Climate Conference in 2007. She clearly announced that the United States was not willing to participate in any agreement to reduce emissions without developing countries taking on more of the burden. When the booing from the remaining delegates subsided, the hostile appeal from the Papua New Guinea spokesman, Kevin Conrad, bluntly declared: "There is an old saying," he said, "If you are not willing to lead then get out of the way." He continued: "We ask for your leadership. We seek your leadership, but if for some reason you are not willing to lead leave it to the rest of us. Please. Get out of the way!" To everyone's surprise and glee, Dobriansky did the unthinkable; she did a 180-degree turn in attitude: "I think that we have come a long way here, and in fact the United States is very committed to this effort, and just wants to really insure that we all will act today. So, Mr. Chairman, let me say to you that we will go forward and join consensus in this today."[3] Rousing the energies of the multitude by assuming a collective voice, Conrad's pithy demand had triggered an unforeseeable change in events. The dominant sign of global power started to stutter as the multitude began to speak on their own terms and for themselves.

Given the focus of sustainability is the sustenance of life, cultural practices committed to reconstituting life necessarily work hard to irritate and challenge any thing or system that would compromise the intricate web supporting life on Earth. That web connects different modalities and ways of life, and in the process life as a whole changes. This could best be described as a system of creative production.[4] It is filled with difference and yet there is no contradiction. As Paul Hawken recognizes in the opening pages of *Blessed Unrest*, it is a source of optimism and is absolutely nonideological. Further, "what binds its constituents is a modus operandi that could be called the autonomy of diversity."[5] It is unorthodox and has no single point of authority dominating the movement of the whole. Hence, when we ask who is in control, it is not a solitary subject who steps forward but a multitude that organizes itself to deterritorialize the social and environmental injustices capitalism and militarism inflict. The multitude constitutes the collective voice and energies that have pushed sustainability culture into the mainstream.[6] In the presence of *potentia* (the power of the multitude), *potestas* (the power of the sovereign) is curbed.

In all its passionate intensity, *potentia* is the state of unimaginableness augmented by and actualized in the force of the multitude. It appears in Conrad's stinging words, which loosened the grip of U.S. power at the Climate Change talks in Bali. It is clearly present amid those defiant individuals whose stained fingers invented a new language through protest by making the whole idea of dispute palpable once more. It vibrates among the bodies shivering together in support of one another and the environment in Tunick's glacial photographs. And it is unexpectedly audible when the lonely tree quietly survives another day inside Didier's structure. *Unimaginableness* differs from *unimaginable* because as an immanent condition it does not aspire to realize what is otherwise impossible—the unimaginable—which would seem to suggest that it merely indulges in the production of imaginary worlds. Rather, the operative mode of unimaginableness is onto-aesthetic. And this is where sustainability culture gallops into the forefront. The aesthetic grammar of culture has the power to distill the otherwise complex language of science into an affective and potent vocabulary—one that vibrates our senses, captures the force of a world undergoing dramatic change, renders visible a reality that is itself not visible, and, more important, germinates a sense of responsibility where previously there was apathy and lethargy at best.

Putting the state of unimaginableness to work, sustainability culture discovers the regenerative force immanent to *potentia*, tapping into its creative and productive energies, all the while daring to imagine and design alternatives to how culture is produced, disseminated, and consumed. Sustainability culture is optimistic insofar as it encourages us to work for a future that is different from the present by focusing more on sociality than profit-maximizing principles. It aspires to create mechanisms that affirm the vitality and dynamic materialism of life as these saturate and perpetuate life. This vital materiality is the ontological energy that all life shares in common. In effect, the aesthetic issue sustainability culture tackles is not one of criticizing the current condition and what once was; it is more a problem of how to invent new values, and its scientific function might best be described as an autoconstituting independence that is produced in dependence with current limits and conditions. These would be the limits that life itself sets and the condition of unimaginableness generating life. Yes, life has limits—Earth's metabolism can gulp down only so much of our waste, and Earth can recycle only a finite amount of the toxins industry spews into the atmosphere. So to claim that sustainability culture is a process of autonomous constitution is merely to suggest that it entails destroying old habits and ways of being (reactive values).

What I have attempted to demonstrate throughout this book is that the political viability of sustainability culture arises when creative thinking and specific experiences are put to work in ways that benefit life equally. Expanding the conditions that bring everyday practices to life, sustainability culture is directly involved with enlarging reality. Providing a theoretical investigation and challenging the habits and conventions that produce social value and systems of violence, it also participates in a process of abstraction. The sovereignty of sustainability culture comes from how it subverts the commodification of the environment and the priorities that accord privilege and power to the few at the expense of the world's poor and life on earth as we know it. As a cultural movement it does not pander to either free market and financialized capitalism (overconsumption) or militarism (irregular warfare). Its qualities encompass a way of knowing and sensing reality that cannot be summarized using the categories of reason, nor can these be evaluated using the criterion of representational truth. Sustainability culture joyfully embraces the ontological force of life in all its creative potential, and in this regard it is without doubt a polemic against the banalization of life. It creates new ways of feeling, thinking, and intuiting life by charting the powers of art and science together, passing through and drawing upon local conditions without ever becoming provincial, gaining independence with every move, and amplifying the productive (as opposed to destructive) force of sociality. Lastly, sustainable considerations that are not cultural are not sustainable. Yet a theory and practice of sustainability that is only cultural ultimately compromises the complexity and dynamism of life.

Notes

Introduction

1. Gro Brundtland, *Our Common Future: The World Commission on Environment and Development* (Oxford: Oxford University Press, 1987).

2. Ibid., xi.

3. Ibid., 1.

4. Fredric Jameson, *Postmodernism: Or the Cultural Logic of Late Capitalism* (Durham: Duke University Press, 1991), 3.

5. Andrew J. Bacevich. *The New American Militarism: How Americans Are Seduced by War* (Oxford: Oxford University Press, 2005).

6. Rem Koolhaas, "Junkspace," in *Harvard School of Design Guide to Shopping*, ed. C. J. Chung (Hong Kong: Taschen, 2002).

7. Slavoj Žižek, *The Parallax View* (Cambridge, MA: MIT Press, 2006), 129.

8. Christian Aid, "The Climate of Poverty: Facts, Fears, and Hopes," *Christian Aid Report*, (May 2006), 12.

9. Judith Butler, "Vikki Bell: Interview with Judith Butler," *Theory, Culture & Society* 16, no. 2 (1999): 165.

Chapter 1

1. Friedrich Nietzsche, *Human, All Too Human: A Book for Free Spirits*, trans. R. J. Hollingdale (Cambridge: Cambridge University Press, 1996), 180.

2. Nietzsche writes: "the will to power is the primitive form of affect, that all other affects are only developments of it; that it is notably enlightening to posit power in place of individual 'happiness' (after which every living thing is supposed to be striving): 'there is a striving for power, for an increase of power';—pleasure is only a symptom of the feeling of power attained, a consciousness of a difference." See

Friedrich Nietzsche, *The Will to Power*, trans. W. Kaufmann and R. J. Hollingdale (New York: Vintage, 1968) #688: 366.

3. Corporate Watch, "Greenwash Fact Sheet" (2001).

4. The Global Climate Coalition finally fell apart in 2002. Cited in Center for Media and Democracy, "Source Watch."

5. Michael Hawthorne, "BP Gets Break on Dumping in Lake," *Chicago Tribune*, July 15, 2007.

6. Bart Stupak, "Stupak BP Hearing Statement," News from Congressman Bart Stupak, September 7, 2006. At the time, Bart Stupak was the Michigan Democrat and Chairman of the House Energy Committees subcommittee.

7. OSHA, "National News Release: USDL 05-1740," September 22, 2005.

8. Ibid.

9. The report also provides a concise history of socially responsible investing in the United States. See Social Investment Forum, *2005 Socially Responsible Investing Trends in the United States*, February 24, 2005, (Washington, DC), 3–4.

10. Ibid. 4.

11. Ibid. 2.

12. Ibid., 1.

13. McDonough, Braungart, Hawken, and Lovins are all avid proponents of the Next Industrial Revolution movement.

14. George Monbiot, *Heat: How to Stop the Planet from Burning* (Cambridge, MA: South End Press, 2007), xvii.

15. Local Authority Pension Fund Forum, "Local Authorities Oppose BP Pay Policy's Lack of Safety Targets," press release, March 26, 2007.

16. Social Investment Forum, *2005 Socially Responsible Investing Trends*, 3.

17. As Deleuze describes it: "In a body the superior or dominant forces are known as *active* and the inferior or dominated forces are known as *reactive*. Active and reactive are precisely the original qualities which express the relation of force with force." See Gilles Deleuze, *Nietzsche and Philosophy*, trans. Hugh Tomlinson (New York: Columbia University Press, 1983), 40.

18. According to statistics compiled by Autodata, sales figures for the Hummer H1 and H2 in July 2002 were 1,922, and in November 2002 this figure escalated to 3,933.

19. Social Investment Forum, *2005 Socially Responsible Investing Trends*, 2.

20. Green Car Congress, "GM's Design Concept HUMMER O_2: Fuel Cell HUMMER That Breathes," July 18, 2007.

21. Green Car Congress. "February 2006 US Hybrid Sales up 44% from Prior Year; Prius down 7.5%," March 7, 2006.

22. Rick Wagoner, "2005/2006 Corporate Responsibility Report," GM Corporate Responsibility.

23. American Marketing Association, "Brand," *Dictionary of Marketing Terms*.

24. See Theodore Levitt. *The Marketing Mode* (New York: McGraw-Hill, 1969).

25. Joseph LePla, Susan Voeller Davis, and Lynne M. Parker. *Brand Driven: The Route to Integrated Branding through Great Leadership* (London: Kogan Page, 2003), 38.

26. Ibid., 41.

27. See Kevin L. Keller, *Strategic Brand Management* (Upper Saddle River, NJ: Prentice Hall, 1998).

28. Jean-Noël Kapferer, *Strategic Brand Management: New Approaches to Creating and Evaluating Brand Equity* (New York: Free Press, 1992), 139.

29. Ian Buchanan, "Space in the Age of Non-Place," *Deleuze and Space*, ed. I. Buchanan and G. Lambert (Edinburgh: Edinburgh UP, 2005), 29.

30. Frederic Jameson, "Future City," *New Left Review* 21 (May/June 2003): 74.

31. Hal Foster, *Design and Crime [and Other Diatribes]* (London: Verso, 2002), 56.

32. Rem Koolhaas, "Junkspace," in *Harvard School of Design Guide to Shopping*, ed. C. J. Chung (Hong Kong: Taschen, 2002).

33. Charles Fishman, *The Wal-Mart Effect* (New York: Penguin, 2006), 3–4, 6.

34. Lee Scott, "Twenty-First Century Leadership," October 24, 2005.

35. Wal-Mart Fact Sheet. "Wal-Mart Opens First Experimental Supercenter," April 10, 2007.

36. Wal-Mart Realty. "Basic List of Available Buildings."

37. Chuihua Judy Chung, ed., *Harvard School of Design Guide to Shopping* (Hong Kong: Taschen, 2002), 51–53, 63.

38. Ibid., 51–53.

39. Ibid., 67.

40. Sze Tsung Leong, ". . . And Then There Was Shopping," *Harvard School of Design Guide to Shopping*, ed. C. J. Chung (Hong Kong: Taschen, 2002), 129.

41. Foster, *Design and Crime*, 23.

42. Fishman, *The Wal-Mart Effect*, 3.

43. Ibid., 6.

44. This idea is influenced by Badiou's concept of the truth event. See Alain Badiou, *Ethics: An Essay on the Understanding of Evil*, trans. P. Hallward (London: Verso, 2001).

45. Foster, *Design and Crime*, 19.

Chapter 2

1. *The Player* (1992) was directed by Robert Altman and stars Tim Robbins. It won a Golden Globe for Best Comedy, and Altman was named "Best Director" at the Cannes Film Festival. The film is a humorous satire of the Hollywood movie industry and tells the story of a Hollywood studio executive who thinks he is being blackmailed by a screenwriter whose script he has rejected.

2. William McIntosh, Rebecca Murray, John Murray, and Debra Sabia, "Are the Liberal Good in Hollywood? Characteristics of Political Figures in Popular Films from 1945 to 1998," *Communications Reports* 16 (2003): 57.

3. Ibid., 65.

4. Ibid.

5. Charles J. Corbett and Richard P. Turco, "Film and Television," Southern California Environmental Report Card 2006 (Los Angeles: University of California, Los Angeles Institute of the Environment, 2006), 10–11.

6. Environmental Media Association, "17th Annual EMA Awards." http://www.ema-online.org. Accessed November 29, 2007.

7. David Ingram, *Green Screen: Environmentalism and Hollywood Cinema* (Exeter, Devon: University of Exeter Press, 2000), 179.

8. Ibid., 180.

9. Ibid.

10. Ibid., 181.

11. The figure estimated by All-Time USA Box Office is $184,208,848. All-Time USA Box Office. http://www.imdb.com/boxoffice/alltimegross. Accessed November 30, 2007.

12. Fredric Jameson, "Metacommentary," *PMLA* 86, no. 1 (1971): 16.

13. Ibid.

14. Jonathon Turley, "Hollywood Isn't Holding Its Lines against the Pentagon," *Los Angeles Times*, August 19, 2003.

15. Aida Hozic, *Hollyworld: Space, Power, and Fantasy in the American Economy* (Ithaca: Cornell University Press, 2001).

16. Ibid., 5.

17. Ibid., 7.

18. Criteria pollutants include nitrogen dioxide (NO_2), carbon monoxide (CO), sulfur dioxide (SO_2), and particulate matter ($PM_{2.5}$ and PM10). Corbett and Turco, "Film and Television," 7.

19. Green house gas emissions include carbon dioxide (CO_2), methane (CH_4), and mitrous oxide (N_2O). Ibid., 8.

20. Tim Flannery explains: "Carbon dioxide is the most abundant of the trace greenhouse gases, and it's produced whenever we burn something and when something decomposes." He provides readers with the following daunting statistics, which are worthwhile quoting at length: "CO_2 is very long lived in the atmosphere: Around 56 percent of all the CO_2 that humans have liberated by burning fossil fuel is still aloft, which is the cause—directly or indirectly—of around 80 percent of all global warming." He adds (and this is where the statistics are the most alarming): "The fact that a known proportion of CO_2 remains in the atmosphere allows us to calculate, in very round numbers, a carbon budget for humanity. Prior to 1800 (the start of the Industrial Revolution), there were about 280 parts per million of CO_2 in the atmosphere, which equates to around 645 gigatons (billion tons) of CO_2 . . . Today the figures are 380 parts per million, or around 869 gigatons. If we wished to stabilize CO_2 emissions at a level double that which existed before the Industrial Revolution . . . we would have to limit all future human emissions to around 660 gigatons. Just over half of this would stay in the atmosphere, raising CO_2 levels to around 1,210 gigatons, or 550 parts per million by 2100." Tim Flannery, *The Weather Makers: How Man Is Changing the Climate and What It Means for Life on Earth* (New York: Atlantic Monthly Press, 2005), 24, 28–29.

21. Corbett and Turco, "Film and Television," 9.

22. Ibid., 10.

23. National Resource Defense Council. "Environmental Achievements of the 79th Annual Academy Awards." http://www.nrdc.org. Accessed July 24, 2007.

24. Leonardo DiCaprio at the 79th Annual Academy Awards, February 25, 2007.

25. Global Green USA, "Global Green Promotes Plug-in Hybrid, All Electric, and Alternative Fuel Vehicles for Oscars Drive." http://www.globalgreen.org/press/releases/2007_2_23_oscarsgreencars.htm. Accessed July 24, 2007.

26. Lary May, *The Big Tomorrow: Hollywood and the Politics of the American Way* (Chicago: Chicago University Press, 2000), 3.

27. Ibid., 1.

28. Neil Postman, *Amusing Ourselves to Death: Public Discourse in the Age of Show Business* (New York: Penguin, 1986), 4.

29. DiCaprio-the-environmentalist may seem like a sudden phenomenon, but his involvement in environmental activism extends at least as far back as 1998, when he founded the Leonardo DiCaprio Foundation. The foundation works with nonprofit and nongovernmental organizations such as Natural Resources Defense Council; Global Green, USA; International Fund for Animal Welfare; National Geographic Kids; Dian Fossey Foundation; Reef Check; Oceana; Santa Monica's Heal the Bay; and U'wa Defense Project. Working in collaboration with the NRDC and Tree Media, DiCaprio has fought hard for the reduction of greenhouse gases and for universal access to clean water. A quick look at the foundation's Web site is filled with information on issues including global warming, biodiversity, and clean water; advice on practical measures for becoming more informed and taking action; and a calendar listing upcoming environmental events. A testimony to his contribution to environmental justice issues, in 2001 Environment Now awarded DiCaprio the Martin Litton Environmental Warrior Award, and in 2003 Global Green USA presented him with the Environmental Leadership Award.

30. *Vanity Fair*, May 2006, 18.

31. After serving as the forty-sixth governor of Texas, George W. Bush became president of the United States of America in 2000, assuming office on January 20, 2001. The fundamental difference between the 1992 Earth Summit's Convention on Global Climate Change that the first President Bush signed and the 1997 Kyoto Protocol that President Bill Clinton signed is that Bush senior's version included voluntary targets and optional guidelines, whereas Clinton agreed to have the reduction of green house gas emissions be mandatory. The Kyoto Protocol aspires to bring together countries from around the world to work on solutions for reducing global warming. Signatories to the Kyoto Protocol agreed to reduce the following greenhouse gas emissions by 5.2 percent (based on the levels present in 1990) when averaged over the period of 2008–2012: carbon dioxide, methane, HFCs, PFCs, sulfur hexafluoride, and nitrous oxide. Countries who are struggling to meet this target are permitted to offset their emissions by trading with countries that have met their own targets.

32. Noah D. Greenwald and Kieran F. Suckling, "Progress or Extinction? A Systematic Review of the U.S. Fish and Wildlife Service's Endangered Species Act Listing Program 1974–2004," Center for Biological Diversity, May 2005, 4.

33. As global warming continues, the main threat to the polar bears is their shrinking habitat. Unlike other species at lower altitudes, polar bears cannot migrate far-

ther north. Other significant threats include global warming, the absence of hunting quotas in Canada and Greenland, poaching in Russia, toxic contamination, shipping, tourism, oil and gas exploration, and development.

34. 2006 IUCN Red List of Threatened Species: Media Package for North America, the World Conservation Union (IUCN), 2006, 1.

35. According to the World Conservation Union (previously known as the International Union for the Conservation of Nature and Natural Resources—IUCN), the survival status of polar bears in 1996 was low risk but conservation dependent. A decade later, they were moved to the red list after a reassessment determined that they were vulnerable. The World Conservation Union, established in October 1948 and headquartered in Gland, Switzerland, is a network that has the support of 83 states, 110 government organizations, and more than 800 nongovernmental organizations, and which includes approximately 10,000 scientists and experts from over 181 countries. Its stated mission is: "to influence, encourage and assist societies throughout the world to conserve the integrity and diversity of nature and to ensure that any use of natural resources is equitable and ecologically sustainable." The World Conservation Union. http://www.iucn.org/en/about. Accessed November 21, 2007.

36. Lee Foote, "NASUSG Calls for Freeze on Polar Bear Reclassification," Sustainable: The Newsletter of the IUCN SSC Sustainable Use Specialist Group, June 2006, 6.

37. Ibid., 7.

38. Ibid.

39. Inuit Circumpolar Council (Canada), "Inuit Oppose and Seek Clarification of IUCN Decision to Change Polar Bear Status to 'Vulnerable,'" ICC press release, May 11, 2006.

40. Inuit Circumpolar Council (Canada), "Inuit Cite IPCC Results as Further Proof of Human Impacts Contributing to Climate Change: Inuit Call on Canada to Recognize Arctic in Foreign Policy and Commit Resources towards Adaptation," ICC press release, February 2, 2007.

41. The Inuit have a long and complex relationship to their environment, and more specifically to polar bears. Traditional folktales tell the story of polar bears helping the Inuit stay alive in their harsh environment by providing the protagonists with food and companionship. The Inuit have a deep respect for the polar bear a position that is reiterated in Inuit mythology that sees no difference between the spirit of a human being and that of other animals.

Chapter 3

1. Savannah Live Well and Prosper, "Data." http://www.seda.org/content.php ?section=data&subsection=population. Accessed October 21, 2007.

2. Don Luymes, "The Fortification of Suburbia: Investigating the Rise of Enclave Communities," *Landscape and Urban Planning*, 39 (1997): 191.

3. Forbes.com. http://www.forbes.com/lists/2005/5/2933.shtml. Accessed August 6, 2007.

4. Savannah Chatham Metropolitan Police, "2006 Neighborhood Crime Statistics," 2006, 3.

5. Ibid.

6. Ibid., 15. See also Daniel Lockwood, "Mapping Crime in Savannah," *Social Science Computer Review* 25, no. 2 (2007): 194–209.

7. Naresh Kumar, "A Spatial Dependency and Causality Analysis of Crime in Savannah, Georgia, 2000," Department of Geography Papers, University of Iowa, 2005.

8. Ann Deehan, "Prevention of Alcohol-Related Crime: Operationalizing Situational and Environmental Strategies," *Crime Prevention and Community Safety: An International Journal* 6, no. 1 (2004): 43–51; K. Graham and S. Wells, "Aggression among Young Adults in the Social Context of the Bar," *Addiction Research* 9 (2001): 193–219; J. Scribner, D. MacKinnon, and J. Dwyer, "The Risk of Assaultive Violence and Alcohol Availability in Los Angeles County," *American Journal of Public Health,* 85, no. 3, 1995: 335–340; L. Zhu, D. M. Gorman, and S. Horel, "Alcohol Outlet Density and Violence: A Geospatial Analysis," *Alcohol and Alcoholism* 39, no. 4, 2004: 369–375.

9. Kumar, "A Spatial Dependency," 3.

10. Ibid., 18.

11. Larry Adelman (producer), "Racial Preferences for Whites: The Houses That Racism Built," Race—the Power of an Illusion, PBS, California Newsreel, 2003.

12. For example, there is another face to the Savannah African-American community, such as that chronicled in the Ralph Mark Gilbert Civil Rights Museum on the corner of Martin Luther King Boulevard and Gaston Street. Here, visitors are introduced to the colorful and proactive character of Black Savannah from slavery to the present day. The museum's displays, oral history testimonies, and interactive exhibits create a picture of this community different from the one on offer by Kumar. Instead of a community suffering from poverty, alcohol consumption, and crime, here is shown the important place religious and cultural institutions hold within the social life of Savannah's African-American community. The social organization around an influential religious institution—Savannah historically has had one of biggest independent black churches in the Old South—has important political implications prompting greater community cohesion. This in turn has fed the characteristic ethos of leadership defining the Savannah African-Americans, a legacy that continues to this day. With a strong cultural and social identity, Savannah's African-American community has a powerful presence in local politics and the cul-

tural life of the city, which was one of the first in the country to integrate African-Americans into the police force. and as of 2007, Since 2004 the office of Chatham County Major has been held by an African-American, Otis Johnson, as have four out of eight City Council seats (two of these are also female).

13. Edward J. Blakely and Mary Gail Snyder, *Fortress America: Gated Communities in the United States* (Washington, DC: Brookings Institution Press, 1997), 1.

14. Ibid., 7.

15. Ibid., vii.

16. Ibid.

17. Ibid.

18. "The Landings on Skidaway Island." http://www.thelandings.com/lifestyle. Accessed August 6, 2007.

19. Robyn Dowling also argues a similar point in her analysis of the neotraditional design principles of New Urbanism. See Robyn Dowling, "Neotraditionalism in the Suburban Landscape: Cultural Geographies of Exclusion in Vancouver, Canada," *Urban Geography* 19, 1998: 105–122.

20. Mike Davis, "Fortress Los Angeles: The Militarization of Urban Space," in *Variations on a Theme Park: The New American City and the End of Public Space*, ed. Michael Sorkin (New York: Noonday Press, 1992), 156.

21. Ian Buchanan, *Fredric Jameson: Live Theory* (London: Continuum, 2006), 98.

22. Karen Till, "Neotraditional Towns and Urban Villages: The Cultural Production of a Geography of Otherness," *Environment and Planning D* 11: 709–732.

23. Buchanan, *Fredric Jameson*, 98.

24. Fredric Jameson, *Postmodernism: Or the Cultural Logic of Late Capitalism* (Durham: Duke University Press, 1999), 38.

25. Arthur W. Bromage and John A. Perkins, "Willow Run Produces Bombers and Intergovernmental Problems," *The American Political Science Review* 36. no. 4 (August 1942), 689–697.

26. See Eeva-Liisa Pelkonen and Donald Albrecht, eds., *Eero Saarinen: Shaping the Future* (New Haven: Yale University Press, 2006).

27. Joseph McCarthy, "Speech on Communists in the State Department," 1950. http://www.civics-online.org/library/formatted/texts/mccarthy.html. Accessed August 10, 2007.

28. Andrew J. Bacevich, *The New American Militarism: How Americans Are Seduced by War* (Oxford: Oxford University Press, 2005), 113.

29. Ibid., 114–115.

30. Jan Martin Bang, *Ecovillages: A Practical Guide to Sustainable Communities* (Gabriola Island, Canada: New Society Publishers, 2005), 18, 102, 140.

31. David W. Orr, *The Nature of Design: Ecology, Culture, and Human Intention* (Oxford: Oxford University Press, 2002), 29.

32. Liz Walker, *Ecovillage at Ithaca: Pioneering a Sustainable Culture* (Gabriola Island, Canada: New Society Publishers, 2005), 86.

33. Bang, *Ecovillages*, 201.

34. Ibid., 130.

35. In the business section of the *New York Times* it was reported that organic farming, although it only constitutes 2.4 percent of the food market, has steadily grown at an annual rate of 15 percent for the past ten years. Clearly, huge profits can be reaped from the organic market, so large corporations are now jumping on the bandwagon. See Melanie Warner, "Wal-Mart Eyes Organic Food," *New York Times*, May 12, 2006.

36. Bang, *Ecovillages*, 136.

37. Orr, *The Nature of Design*, 31.

Chapter 4

1. The White House Historical Association. http://www.whitehousehistory.org/02/subs/02_b.html. Accessed September 21, 2007.

2. Otherwise referred to as the Eisenhower Executive Office Building.

3. Ted Shelton, "Greening the White House: Executive Mansion as Symbol of Sustainability," *Journal of Architectural Education* 60, no. 4, May 2007: 31.

4. William D. Browning, Dianna Lopez Barnett, Mark Ginsberg, and Anne Sprunt-Crawley. "Greening the White House: A Comprehensive Energy and Environmental Retrofit," 5. http://www.rmi.org/images/PDFs/BuildingsLand/D94-09_GrnWhiteHouse.pdf. Accessed September 19, 2007.

5. Shelton, "Greening the White House," 31.

6. Simply put, a frame integrates a group of elements chosen on the basis of perceived similarity—symmetrical organization extending from the building throughout the grounds, classical features that stylistically correlate U.S. democratic roots with those of the Ancient Greeks, a soft green lawn that brings the solid white stone into sharp relief, all of which are secured behind an impermeable iron fence.

7. Jimmy Carter, "Energy Policy," April 18, 1977. http://www.pbs.org/wgbh/amex/carter/filmmore/ps_energy.html. Accessed September 19, 2007.

8. Brink Lindsey, *The Age of Abundance: How Prosperity Transformed America's Politics and Culture* (New York: Harper Collins, 2007), 245.

9. Jimmy Carter, "The Crisis of Confidence," July 15, 1979. http://www.pbs.org/wgbh/amex/carter/filmmore/ps_crisis.html. Accessed September 20, 2007.

10. Carter, "The Crisis of Confidence."

11. Jimmy Carter, "Energy Policy," April 18, 1977. http://www.pbs.org/wgbh/amex/carter/filmmore/ps_energy.html. Accessed September 19, 2007.

12. Carter, "The Crisis of Confidence."

13. At the time Carter delivered his "Crisis of Confidence" speech, the United States was importing 43 percent of its annual oil supplies. See Bacevich, *The New American Militarism*, 102.

14. It is interesting to note that Carter assigned himself the role of First Preacher. Ibid., 105.

15. Ibid.

16. Giorgio Agamben, *Homo Sacer: Sovereign Power and Bare Life*, trans. Daniel Heller-Roazen (Stanford, CA: Stanford University Press, 1998), 18.

17. Carter, "The Crisis of Confidence."

18. James Davison Hunter, *Culture Wars: The Struggle to Define America* (New York: Basic Books, 1991).

19. Lindsey, *The Age of Abundance*, 239.

20. The White House, The Greening of the White House: Six Year Report, November 1999, 5.

21. Shelton, "Greening the White House," 33.

22. Ronald Reagan Obituary. *New York Times*, June 7, 2004. http://query.nytimes.com/gst/fullpage.html?res=9D01E4DE1E31F934A35755C0A9629C8B63. Accessed September 27, 2007.

23. Bacevich, *The New American Militarism*, 108.

24. Ibid., 105.

25. Ibid.

26. Ibid., 107–108.

27. Bill Clinton, "Reaffirming the U.S. Commitment to Protect the Global Environment," April 26, 1993.

28. Shelton, "Greening of the White House," 35.

29. The White House, The Greening of the White House, 6.

30. Ibid., 1.

31. Richard F. Grimmett, "Instances of Use of United States Armed Forces Abroad, 1798–1995," CRS Report 96-119F, February 6, 1996, 18–25.

32. Clinton, "Reaffirming the U.S. Commitment."

33. Ibid.

34. John M. Broder, "At Climate Meeting Bush Does Not Specify Goals," *New York Times*, September 29, 2007. http://www.nytimes.com/2007/09/29/washington/29climate.html?ref=americas. Accessed September 29, 2007.

35. Ecorazzi, "President Bush Texas Ranch is Off-Grid, Eco-Friendly!" http://www.ecorazzi.com/?p=1601. Accessed September 29, 2007.

36. Nick Rosen, "Meanwhile, Back at the Ranch," *Off-Grid*, February 18, 2007. http://www.ecorazzi.com/?p=1601. Accessed September 29, 2007.

37. Brian Faler, "White House Turns Up Heat with Solar Energy at Spa," *Washington Post*, Monday February 3, 2003: A21.

Chapter 5

1. U.S. Army, "U.S. Army Sustainability—Goals." http://www.sustainability.army.mil/overview/goals.cfm. Accessed June 29, 2007.

2. Stephen I. Schwarz cited in Robert F. Durant, *The Greening of the U.S. Military: Environmental Policy, National Security, and Organizational Change* (Washington DC: Georgetown University Press, 2007), 1.

3. Samuel P. Huntington, *The Soldier and the State: The Theory and Politics of Civil-Military Relations* (Cambridge, MA: Belknap Press, 1981), 11.

4. Naomi Klein, *The Shock Doctrine: The Rise of Disaster Capitalism* (New York: Metropolitan Books, 2007).

5. Braden Allenby, "New Priorities in U.S. Foreign Policy: Defining and Implementing Environmental Security," in *The Environment, U.S. International Relations, and Foreign Policy*, ed. Paul G. Harris (Washington, DC: Georgetown University Press, 2001), 45.

6. Andrew J. Bacevich, *The New American Militarism: How Americans Are Seduced by War* (Oxford: Oxford University Press, 2005), 25.

7. U.S. Army, "U.S. Army Sustainability—Environmental Management Systems." http://www.sustainability.army.mil/tools/programtools_ems.cfm. Accessed November 14, 2007

8. As Anthony Lake stated: "The second feature of this era is that we are its dominant power. Those who say otherwise sell America short. The fact is, we have the world's strongest military, its largest economy and its most dynamic, multiethnic society. We are setting a global example in our efforts to reinvent our democratic and market institutions. Our leadership is sought and respected in every corner of the world. As Secretary Christopher noted yesterday that is why the parties to last week's dramatic events chose to shake hands in Washington. Around the world, America's power, authority and example provide unparalleled opportunities to lead." Anthony Lake, "From Containment to Enlargement," delivered at Johns Hopkins University September 21, 1993. http://www.mtholyoke.edu/acad/intrel/lake doc.html. Accessed November 14, 2007

9. Cited in Durant, *The Greening of the U.S. Military*, 30.

10. Bill Clinton, "First Inaugural," January 20, 1993. http://www.millercenter.vir ginia.edu/scripps/digitalarchive/speeches/clinton/wjc_1993_0120. Accessed November 15, 2007.

11. Jon Barnett, "Environmental Security and U.S. Foreign Policy: A Critical Examination," in *The Environment*, ed. Harris, 74–75.

12. See Dieter Koenig, "Sustainable Development: Linking Global Environmental Change to Technology Cooperation," in *Environmental Policies in the Third World: A Comparative Analysis*, ed. O. P. Dwivedi and Dhirendra K. Vajpeyi (Westport, CT: Greenwood Press, 1995), 13–14.

13. Barnett, "Environmental Security," 75

14. Durant, *The Greening of the U.S. Military*, 7.

15. Ibid., 10.

16. Lake, "From Containment to Enlargement."

17. Lake specifically stated: "Ultimately, it is through our support for democracy and sustainable development that we best enhance the dramatic new winds of change that are stirring much of the developing world." Ibid.

18. Cited in Durant, *The Greening of the U.S. Military*, 59.

19. The United States Commission on National Security/21st Century, Road Map for National Security: Imperative for Change, February 15, 2001.

20. In the foreword, the problem was stated in terms of how "American principles, interests and national purpose" could be supported. Ibid., iv.

21. Ibid., v.

22. The effect of such ecogeopolitical discourse is that the environment is understood as an external entity—one that is easily turned into a military resource, then

manipulated and managed for the purposes of U.S. power and interest. Cited in Barnett, "Environmental Security," 75.

23. U.S. Army, "2007 Posture Statement." http://www.army.mil/aps/07. Accessed June 29, 2007.

24. FY 2007 Department of Defense Budget, February 6, 2006. http://www.defenselink.mil/news/Feb2006/d20060206slides.pdf. Accessed June 29, 2007.

25. Robert Higgs, "The Trillion Dollar Defense Budget Is Already Here," Independent Institute, March 15, 2007. http://www.independent.org/newsroom/article.asp?id=1941. Accessed June 29, 2007.

26. U.S. Census Bureau, "Poverty." http://www.census.gov/hhes/www/poverty/poverty06/pov06hi.html. Accessed November 18, 2007.

27. Bacevich, *The New American Militarism*, 1.

28. Stockholm International Peace Research Institute (SIPRI), *SIPRI Yearbook 2007: Armaments, Disarmament and International Security* (Oxford: Oxford University Press, 2007), 269.

29. These statistics are for 2006. Care International, "Children and Poverty Campaign." http://www.care.org/campaigns/childrenpoverty/index.asp?source=1707402 50000&WT.srch=1. Accessed November 18, 2007.

30. Charles Cater, "The Political Economy of Conflict and UN Intervention: Rethinking the Critical Cases of Africa," in *The Political Economy of Armed Conflict: Beyond Green and Grievance*, eds. Karen Ballentine and Jake Sherman (Boulder, CO: Lynne Rienner Publishers, 2003), 19–45.

31. Amy Goodman, "Exclusive: Facing Seven Years in Jail, Environmental Activist Daniel McGowan Speaks Out about the Earth Liberation Front, the Green Scare, and the Government's Treatment of Activists As 'Terrorists,'" Democracy Now, June 11, 2007. http://www.democracynow.org/article.pl?sid=07/06/11/142258. Accessed November 17, 2007.

32. Founded in Brighton, England, in 1994, the Earth Liberation Front is an underground movement that aims to use "direct action to sabotage corporations and government agencies that profit from the destruction of the natural environment." See "Target Opportunity: Earth Liberation Front." http://www.targetofopportunity.com/elf.htm. Accessed November 18, 2007.

33. At the time of his arrest McGowan was working for WomensLaw.org, an agency that assists women who are trying to leave abusive partners, and he was enrolled in the Tri-State College of Acupuncture.

34. The eleven indicted were: Joseph Dibee, Chelsea Dawn Gerlach, Sarah Kendall Harvey, Daniel Gerard McGowan, Stanislas Gregory Meyerhoff, Josephine Sunshine

Overaker, Jonathan Mark Christopher Paul, Rebecca Rubin, Suzanne Savoie, Darren Todd Thurston, and Kevin M. Tubbs. See CNN.com Law Center. "Eleven Indicted on Ecoterror Charges," January 20, 2006. http://www.cnn.com/2006/LAW/01/20/ecoterror.indictments/index.html. Accessed November 18, 2007.

35. These figures were cited consistently throughout the media in late 2006. For example, see BBC News, "Huge Rise in Iraqi Death Tolls," October 11, 2006. http://news.bbc.co.uk/2/hi/middle_east/6040054.stm. Accessed November 20, 2007.

36. The Civil Liberties Defense Center provides legal protection to environmental and social-justice activists against corporate and governmental violence against civil liberties. See Goodman, "Exclusive," Democracy Now.

37. Jacques Ranciére, *Hatred of Democracy*, trans. Steve Concoran (London: Verso, 2006).

38. Ibid., 5.

39. Ibid., 6.

40. Sabrina Tavernise, 'Iraqi Death Toll Exceeds 34,000in '06, UN Says', *International Herald Tribune*, January 16, 2007.

41. Here I would agree wholeheartedly with Monbiot, who believes that the military operatpions of countries like the United Kingdom—and I would add America and Australia into the mix—should be confined to peacekeeping: "as much for the sake of the environment as for the sake of public finance and world peace." George Monbiot, *Heat: How to Stop the Planet from Burning* (Cambridge, MA: South End Press, 2007), 60.

42. Ari Fleischer, "Statement by the Press Secretary on the Geneva Convention," The White House, May 7, 2003. http://www.whitehouse.gov/news/releases/2003/05/20030507-18.html. Accessed December 31, 2007.

Chapter 6

1. Lawrence Summers issued this memorandum December 12, 1991, and it became public in February 1992. Summers was the chief economist of the World Bank and U.S. Treasury Secretary under the Clinton Administration. The full contents of the memo were as follows:

"Dirty" Industries: Just between you and me, shouldn't the World Bank be encouraging MORE migration of the dirty industries to the LDCs [Less Developed Countries]? I can think of three reasons:

1. The measurements of the costs of health impairing pollution depends on the foregone earnings from increased morbidity and mortality. From this point of view a given amount of health impairing pollution should be done in the country with the lowest cost, which will be the country with the lowest wages. I think the economic logic behind dumping a load of toxic waste in the lowest wage country is impeccable and we should face up to that.

2. The costs of pollution are likely to be nonlinear, as the initial increments of pollution probably have very low cost. I've always though that under-populated countries in Africa are vastly
UNDER-polluted, their air quality is probably vastly inefficiently low compared to Los Angeles
or Mexico City. Only the lamentable facts that so much pollution is generated by nontradable
industries (transport, electrical generation) and that the unit transport costs of solid waste are so
high prevent world welfare enhancing trade in air pollution and waste.

3. The demand for a clean environment for aesthetic and health reasons is likely to have very
high income elasticity. The concern over an agent that causes a one in a million change in the
odds of prostrate cancer is obviously going to be much higher in a country where people survive
to get prostrate cancer than in a country where under 5 mortality is 200 per thousand. Also,
much of the concern over industrial atmosphere discharge is about visibility impairing particu-
lates. These discharges may have very little direct health impact. Clearly trade in goods that
embody aesthetic pollution concerns could be welfare enhancing. While production is mobile
the consumption of pretty air is a nontradable.

The problem with the arguments against all of these proposals for more pollution in LDCs
(intrinsic rights to certain goods, moral reasons, social concerns, lack of adequate markets, etc.)
could be turned around and used more or less effectively against every Bank proposal for
liberalization.

http://www.whirledbank.org/ourwords/summers.html. Accessed September 29,
2007.

2. William McDonough and Michael Braungart, *Crade-to-Cradle* (New York: Northpoint Press, 2002), 55.

3. Heather Rogers, *Gone Tomorrow: The Hidden Life of Garbage* (New York: The New
Press, 2005), 6.

4. Ibid., 2.

5. U.S. Environmental Protection Agency, "Recycling." http://www.epa.gov/epao
swer/non-hw/muncpl/recycle.htm. Accessed August 23, 2007.

6. Rogers, *Gone Tomorrow*, 187.

7. Ibid., 183–205.

8. Ibid., 185.

9. *The Michigan Daily*, "Legislation Seeks Ban to Garbage Imports," February
24, 1997. http://www.pub.umich.edu/daily/1997/feb/02-24-97/news/news10.html.
Accessed August 23, 2007.

10. Ohio Environmental Protection Agency, "2004 Out-of-State Waste," Ohio EPA
Fact Sheet, November 2005, 2.

11. Ibid.

12. For more on the phenomenon on postindustrialization see Daniel Bell, *The
Coming of Post-Industrial Society: A Venture into Social Forecasting* (New York: Basic
Books, 1973).

13. Michael Moore in *Roger and Me* (1989), directed by Michael Moore.

14. Jennifer Clapp, "Seeping through the Regulatory Cracks," *SAIS Review* XXII, no. 1 (2002): 151.

15. Ibid.

16. Nisha Thakker, "India's Toxic Landfills: A Dumping Ground for the World's Electronic Waste," *Sustainable Development Law and Policy* (Spring 2006): 58.

17. Clapp, "Seeping through the Regulatory Cracks," 141.

18. Basel Action Network, "Exporting Harm: The Hightech Trashing of Asia," press release, October 24, 2005.

19. Basel Action Network and Silicon Valley Toxics Coalition, "High Tech Toxic Trash from USA Found to Be Flooding Asia," press release, October 24, 2005.

20. Thakker, "India's Toxic Landfills," 60.

21. Ibid.

22. Basel Convention on the Control of Transboundary Movements of Hazardous Wastes and Their Disposal, 2. http://www.basel.int/text/con-e-rev.pdf. Accessed June 6, 2008.

23. Ibid., 6

24. Ibid., 9.

25. McDonough and Braungart, *Cradle-to-Cradle*, 56.

26. Ibid.

27. Rogers, *Gone Tomorrow*, 3.

28. Julia Kristeva, *Powers of Horror: An Essay on the Understanding of Abjection*, trans. Leon S. Roudiez (New York: Columbia University Press, 1982), 4.

29. Susan Strasser, *Waste and Want: A Social History of Trash* (New York: Holt, 2000), 286.

30. Ibid.

31. Mary Douglas, *Purity and Danger: An Analysis of Concepts of Pollution and Taboo* (London: Routledge, 1966), 36.

32. Ibid., 35–36.

33. Strasser, *Waste and Want*, 275.

34. Ibid., 281.

35. Ibid., 283.

36. Malcolm G. Scully, "Making Peace with Diversity," *Chronicle of Higher Education* 49, no. 29: B15. See also, Michael L. Rosenzweig, *Win-Win Ecology: How the Earth's Species Can Survive in the Midst of Human Enterprise* (Oxford: Oxford University Press, 2003).

37. John Bellamy Foster, *Marx's Ecology: Materialism and Nature* (New York: Monthly Review Press, 2000).

38. Ibid., 16.

39. Cited in John L. Eliot, "A Dump Reviled, Revered," *National Geographic* 203, no. 1 (January 2003): 9.

40. William Young, "A Dump No More," *American Forests* 101 (August 1995): 58.

41. McDonough and Braungart, *Crade-to-Cradle*, 33.

42. New Yorkers may have gained new parklands, but New Jersey, Pennsylvania, and Virginia are not so fortunate. The long-term plan is to use a barge that will haul on a daily basis 6,500 tons of New York's waste to a New Jersey transfer station; this will then be transported by rail or barge out of state.

Chapter 7

1. Christian Aid, "The Climate of Poverty: Facts, Fears and Hopes," *Christian Aid Report* (May 2006), 12. http://www.christianaid.org.uk/Images/climate_of_poverty_tcm15-21613.pdf. Accessed December 31, 2007.

2. United Nations High Commission for Refugees (UNHCR), *UNHCR Handbook for Emergencies*, 2nd ed. (2000), 144.

3. Tom Corsellis and Antonella Vitale. *Transitional Settlement, Displaced Populations*, Oxfam (Cambridge: University of Cambridge, 2005), 6.

4. Nandan interviewed by K. P. Sasi and Max Martin, "Shelter: Little Problems Need Big Attention," Tsunami Response Watch. http://www.tsunamiresponsewatch.org/2006/11/02/shelter-little-problems-need-big-attention/#more-724. Accessed August 17, 2007.

5. Ibid.

6. Charles Correa, *Housing and Urbanization: Building Solutions for People and Cities* (London: Thames and Hudson, 2000); Paul Oliver, *Built to Meet Needs: Cultural Issues in Vernacular Architecture* (New York: Architectural Press, 2006); Paul Oliver, *Dwellings: The Vernacular House Worldwide* (London: Phaidon, 2007); and Geoffrey Payne and Michael Majale, *The Urban Housing Manual: Making Regulatory Frameworks for the Poor* (London: Earthscan, 2004).

7. Kofi Annan, "International Strategy for Disaster Reduction," October 11, 2006. http://www.unisdr.org/eng/public_aware/world_camp/2006-2007/iddr/2006-iddr .htm. Accessed Sunday, May 20, 2007.

8. United Nations, "Press Conference by United Nations Deputy Emergency Relief Coordinator on Recent Floods in South Asia," August 9, 2007. http://www.un.org/ News/briefings/docs/2007/070809_Wahlstrom.doc.htm. Accessed August 13, 2007.

9. Tim Flannery, *The Weather Makers: How Man Is Changing the Climate and What It Means for Life on Earth* (New York: Atlantic Monthly Press, 2005), 144.

10. Ibid., 143.

11. Christian Aid, "The Climate of Poverty," 11.

12. Flannery, *The Weather Makers*, 138.

13. Theodor W. Adorno, *Probleme der Moralphilosophie*, 30. Cited in Kritik der ethischen Gewalt: Adorno Lectures, 2002, Judith Butler (Frankfurt: Institut für Sozialforschung an der Johann Wolfgang Goethe-Universität, 2003), 13.

14. Theodor W. Adorno, *Aesthetic Theory*, trans. Robert Hullot-Kentor (Minneapolis: University of Minnesota Press, 1998).

15. Architecture for Humanity, ed., *Design Like You Give a Damn: Architectural Responses to Humanitarian Crises* (New York: Metropolis Books, 2006), 98.

16. I have discussed this issue at length elsewhere. See Adrian Parr, *Memorial Culture: Desire, Singular Memory and the Politics of Trauma* (Edinburgh: Edinburgh University Press, 2008).

17. Judith Butler, *Bodies That Matter: On the Discursive Limits of Sex* (London: Routledge, 1993), 29.

18. Corsellis and Vitale. *Transitional Settlement*, 21.

19. Oxfam International, "The Tsunami's Impact on Women," Oxfam Briefing Note, March 2005, 6. http://www.oxfam.org.uk/what_we_do/issues/conflict _disasters/downloads/bn_tsunami_women.pdf. Accessed December 27, 2007.

20. Ibid., 2.

21. Gayatri Chakravorty Spivak, "The Intervention Interview, with Francess Bartkowski," *The Postcolonial Critic: Interviews, Strategies, Dialogues*, ed. Sarah Harasym (London: Routledge, 1990), 120.

22. Sarah Vaill, *The Calm in the Storm: Women Leaders in Gulf Coast Recovery* (Women's Funding Network, 2006), 3; Erica Williams, Olga Sorokina, Avis Jones-DeWeever, and Heidi Hartmann, "The Women of New Orleans and the Gulf Coast: Multiple Disadvantages of Key Assets for Recovery, Part II—Gender, Race, and Class

in the Labor Market," briefing paper, Institute for Women's Policy Research, August 2006, 1.

23. Vaill, *The Calm in the Storm*, 2.

24. Williams et. al., "The Women of New Orleans and the Gulf Coast," 2.

25. Ibid., 3.

26. Ibid.

27. Vaill, *The Calm in the Storm*, 3.

28. Ibid., 4.

29. Williams et. al., "The Women of New Orleans and the Gulf Coast," 12.

30. Ibid., 6.

31. Vaill, *The Calm in the Storm*, 3.

32. The Institute for Women's Policy Research (IWPR) suggested: "A number of policies could enable women in the area to get back on their feet, overcome the structural barriers to good jobs and good wages, and attain an improved standard of living, and would ensure that women's talents are utilized to the fullest in rebuilding the region." Williams et. al., "The Women of New Orleans and the Gulf Coast," 20.

33. "Briefing," *Architectural Record* 01 (2008): 143.

34. Christian Aid, "The Climate of Poverty," 7.

35. Flannery, *The Weather Makers*, 137.

36. Ibid.

37. Ibid., 146.

38. Ibid., 149.

Chapter 8

1. UN-HABITAT, *State of the World's Cities, 2006–2007: The Millennium Development Goals and Urban Sustainability* (Kenya: Earthscan, 2006), iii.

2. Mike Davis, *Planet of the Slums* (London: Verso, 2006), 17.

3. Fabio Soares and Yuri Soares, *The Socio-Economic Impact of Favela-Bairro: What Do the Data Say?* (Washington, DC: Inter-American Development Bank, August 2005) 13.

4. Ibid., 13.

5. UN-HABITAT, *State of the World's Cities*, 164–165.

6. Soares and Soares, *The Socio-Economic Impact of Favela-Bairro*, 2.

7. Janice. E. Perlman, "The Myth of Marginality Revisited: The Case of Favelas in Rio de Janeiro, 1969–2003," Urban Research Symposium 2005, The World Bank Group, March 24, 2005: 3. http://www.worldbank.org/urban/symposium2005/papers/perlman.pdf. Accessed April 19, 2007.

8. Soares and Soares, *The Socio-Economic Impact of Favela-Bairro*, 5.

9. In 1964 Brazil entered into a military dictatorship. In 1971 the ruling authorities decided to implement a set of plans aimed at completely eradicating the favelas within five years and resettling the residents in welfare housing schemes farther away from the city. The selection criteria used to decide which favelas would be relocated first was far from interest-free. Of the 16,467 favela houses demolished between 1968 and 1972, 59.5 percent of these were in the wealthy area of Zona Sul. Meanwhile, the overall favela population continued to escalate, and by 1980 more than 700,000 residents were living in favelas; only 1 percent of these had public sewerage, 6 percent had a water system, and 17 percent had trash collection. Throughout the 1980s the favela population in Rio alone grew 50.7 percent, leaving the city with 3.64 million poor. See Jorge Fiori, Liz Riley, and Ronaldo Ramirez, "Favela-Bairro and a New Generation of Housing Programmes for the Urban Poor," *Geoforum* 32, no. 4 (November 2001), 8–10. http://www.sciencedirect.com. Accessed April 15, 2007.

10. Rodolfo, Machado, "Memoir of a Visit," in *The Favela-Bairro Project*, ed. Rodolfo Machado (Cambridge, MA: Harvard Design School, 2003), 11.

11. Ibid., 15.

12. Brooke Hodge, "The Favela-Bairro Project" in in *The Favela-Bairro Project*, ed. Machado, 19.

13. The work of Favela-Bairro received the esteemed Habitat Award from the United Nations.

14. Paulo Bótas, "Rio de Janeiro to spend US$1 Billion on Innovative Slum Improvement Programme," City Mayors Environment, December 5, 2003. http://www.citymayors.com/report/rio_favelas.html. Accessed April 26, 2007.

15. Judith Butler, *Bodies That Matter: On the Discursive Limits of Sex* (London: Routledge, 1993), xi.

16. Soares and Soares, *The Socio-Economic Impact of Favela-Bairro*, 11.

17. Ibid., 12.

18. Ibid., 8.

19. Fiori, Riley, and Ramirez, "Favela Bairro," 24.

20. UN-HABITAT, *State of the World's Cities*, 172.

21. Jorge Fiori, Liz Riley, and Ronaldo Ramirez. *Urban Poverty Alleviation through Environmental Upgrading in Rio De Janeiro: Favela Bairro*. R7343. Draft Research Report. (Development Planning Unit at the University College London, March 2000), 132.

22. UN-HABITAT, *State of the World's Cities*.

23. Fiori, Riley, and Ramirez. *Urban Poverty Alleviation*, 20.

24. Davis, *Planet of the Slums*, 25; Perlman, "The Myth of Marginality Revisted," 1.

25. Perlman, "The Myth of Marginality Revisited," 20.

26. World Bank Group, "Tracing the Evolution of Poverty in Rio's Favelas." http://lnweb18.worldbank.org/LAC/LAC.nsf/ECADocbyUnid/C9413A90E66702C085256F9A004B79F8?Opendocument. Accessed October 10, 2007.

27. Perlman, "The Myth of Marginality Revisited," 15n33.

28. Soares and Soares, "The Socio-Economic Impact of Favela-Bairro," 7.

29. World Bank Group, "Tracing the Evolution of Poverty in Rio's Favelas"; Janice. E. Perlman, *The Myth of Marginality: Urban Poverty and Politics in Rio de Janeiro* (Berkeley: University of California Press, 1980).

30. Janice. E. Perlman, "Longitudinal Panel Studies in Squatter Communities: Lessons from a Re-study of Rio's Favelas: 1969–2003," Joint World Bank Workshop Urban Longitudinal Research Methodology. London, May 28–29, 2003, 9.

31. Ibid.

32. This analysis uses a point Fredric Jameson makes in his book *The Political Unconscious* (Ithaca, NY: Cornell University Press, 1981), 88.

33. Thomas Barnett, *The Pentagon's New Map: War and Peace in the Twenty-First Century* (New York: Putnam, 2004).

34. Hernando De Soto, *The Mystery of Capital: Why Capitalism Triumphs in the West and Fails Everywhere Else* (New York: Basic Books, 2003).

35. Perlman, "The Myth of Marginality Revisted," 1.

36. Robert Neuwirth, *Shadow Cities: A Billion Squatters, a New Urban World* (New York: Routledge, 2006), 17.

37. Deleuze and Guattari write: "Total war is not only a war of annihilation but arises when annihilation takes as its 'center' not only the enemy army, or the enemy State, but the entire population and its economy." Gilles Deleuze and Félix Guattari, *A Thousand Plateaus: Capitalism and Schizophrenia*, trans. Brian Massumi (London: Athlone Press, 1987), 421.

38. As Deleuze and Guattari write, they think the machinic, "relates not to the production of goods but rather to a precise state of intermingling of bodies in a society,

including all the attractions and repulsions, sympathies and antipathies, altera-tions, amalgamations, penetrations, and expansions that affect bodies of all kinds in their relations to one another." See Deleuze and Guattari. *A Thousand Plateaus*, 90.

39. Douglas Kelbaugh, preface to *Everyday Urbanism*, ed. Rahul Mehrotra (Ann Arbor, MI: University of Michigan, 2005), 8.

40. Douglas Kelbaugh, preface to *Post Urbanism and ReUrbanism*, ed. Roy Strickland (Ann Arbor, MI: University of Michigan, 2005), 8.

41. Steven Petersen, "Urban Design in Lower Manhattan," in *Post Urbanism and ReUrbanism*, ed. Strickland, 28.

42. Margaret Crawford in *Everyday Urbanism*, ed. Rahul Mehrotra 32.

43. Stephen Luoni is chair of Architecture and Urban Studies at the University of Arkansas Community Design Center. http://uacdc.uark.edu.

44. Cited in John Bellamy Foster, *Marx's Ecology: Materialism and Nature* (New York: Monthly Review Press, 2000), 200.

45. The Viva Favela Web site also hosts the local radio station—Rede Viva Favela—that connects the favela community to the rest of Rio.

46. Rejane Reis, "Teaching Tourism in the Slums," Favela Rocinha Tourism Work-shop. http://www.favelatourismworkshop.com. Accessed April 24, 2007.

47. For more information on these programs visit the COAV Web site. http://www .coav.org.br.

48. Judith Butler, "Vikki Bell: Interview with Judith Butler," *Theory, Culture & Society* 16, no. 2 (1999): 166.

49. Ibid.

Chapter 9

1. Gregory Mock, "How Much Do We Consume?" World Resources 2000–2001, June 2000. http://earthtrends.wri.org/features/view_feature.php?theme=6&fid=7. Accessed December 23, 2007.

2. Ibid.

3. Tim Flannery, *The Weather Makers: How Man Is Changing the Climate and What It Means for Life on Earth* (New York: Atlantic Monthly Press, 2005), 59.

4. Intergovernmental Panel on Climate Change Fourth Assessment Report Working Panel. *Climate Change 2007: Synthesis Report* (IPCC), Topic 1: 1. http://www.ipcc.ch/ ipccreports/ar4-syr.htm. Accessed December 25, 2007.

5. Jeffrey Sachs, *The End of Poverty: Economic Possibilities for Our Time* (New York: Penguin, 2005), 19.

6. George Monbiot, "Environmental Feedback: A Reply to Clive Hamilton," *New Left Review* 45 (May/June 2007): 107.

7. Ibid., 112.

8. A classic example of a service-based approach to sustainable design would be the Portland, Oregon, car-sharing company Flexcar (now Zipcar), whereby members pay an annual fee and choose a rate plan that suits their automobile needs (insurance, gasoline, vehicle maintenance, and cleaning), scheduling pick-up and return times for vehicles as required.

9. William, McDonough and Michael Braungart, *Cradle-to-Cradle: Remaking the Way We Make Things* (New York: Northpoint Press, 2002), 91.

10. Andres R. Edwards, *The Sustainability Revolution: Portrait of a Paradigm Shift* (Philadelphia, PA: New Society Publishers, 2005), 99. McDonough and Braungart. *Cradle-to-Cradle*, 67.

11. Ibid., 62.

12. Performance artist Joseph Beuys defined this approach to art, ontology, and the environment with this formula: Creativity = Capital.

13. Muhammad Yunus, *Creating a World without Poverty* (New York: Public Affairs, 2007), 6.

14. Ibid., 75.

15. Christian Aid, "The Climate of Poverty: Facts, Fears, and Hopes," *Christian Aid Report* (May 2006). http://www.christianaid.org.uk/Images/climate_of_poverty_tcm15-21613.pdf. Accessed December 31, 2007.

16. The 2015 eight MDGs for the world's poorest are: (1) Eradicate extreme poverty and hunger; (2) Achieve universal primary education; (3) Promote gender equality and empower women; (4) Reduce child mortality; (5) Improve maternal health; (6) Combat HIV/AIDS, malaria, and other diseases; (7) Ensure environmental sustainability; and (8) Develop a global partnership for development.

17. Christian Aid, "The Climate of Poverty," 2.

18. Ibid.

19. "The PlayPump Water System," *PlayPumps International*. http://www.playpumps.org/site/c.hqLNIXOEKrF/b.2589561/k.C08/The_PlayPump_System__The_Water_Problem.htm. Accessed June 9, 2008. Barbara Frost. "Water and Sanitation: The Silent Emergency," *UN Chronicle* xlv, no. 1, 2008. http://www.un.org/Pubs/chronicle/2007/issue4/0407p87.html. Accessed June 9, 2008.

20. McDonough Braungart Design Chemistry, "About MBDC." http://www .mcdonough.com/product.htm. Accessed December 26, 2007.

21. One Laptop Per Child, "Vision." http://laptop.org/vision/index.shtml. Accessed December 26, 2007.

22. Steve Stecklow, "A Little Laptop with Big Ambitions." *Wall Street Journal*, November 24, 2007, A1.

23. Architecture for Humanity, ed., *Design Like You Give a Damn: Architectural Responses to Humanitarian Crises* (New York: Metropolis Books, 2006), 31.

24. McDonough and Braungart, *Cradle-to-Cradle*, 90.

25. Vandana Shiva, *Biopiracy: The Plunder of Nature and Knowledge* (Cambridge, MA: South End Press, 1997), 7.

26. Ibid.

27. For more on the issue of biopiracy see Shiva, *Biopiracy*.

28. Architecture for Humanity, "Bay View Model Block Project." http://www.arch itectureforhumanity.org/?q=node/165. Accessed December 26, 2007.

29. BBC News. "Q&A: The U.S. and Climate Change," February 14, 2002. http:// news.bbc.co.uk/2/hi/americas/1820523.stm. Accessed November 1, 2007.

30. Roger Rosenblatt and San Bruno. "The Man Who Wants Buildings to Love Kids," *Time*, February 15, 1999. http://www.time.com/time/reports/environment/ heroes/heroesgallery/0,2967,mcdonough2,00.html. Accessed November 1, 2007.

31. Sachs, *The End of Poverty*, 20.

32. Ibid., 267.

33. Arundhati Roy, *The Algebra of Infinite Justice* (London: Flamingo, 2002), 15.

34. Sachs, *The End of Poverty*, 288.

Conclusion

1. The image was used within an article discussing the IPCC meeting, and the caption that accompanied it read: "As the polar bear's domain shrinks inexorably, two make the best of things on a chunk of ice off Northern Alaska." Oliver Burkeman, "The Warm-Mongers Rest Their Case," *The Age*, February 4, 2007.

2. Cited in Paul Hawken, *Blessed Unrest: How the Largest Movement in the World Came into Being and Why No One Saw It Coming* (New York: Viking, 2007), 24.

3. Richard Harris, "Climate Roadmap Emerges from Grueling Bali Talks," National Public Radio, December 15, 2007. http://www.npr.org/templates/story/story.php ?storyId=17519540. Accessed January 16, 2008.

4. My book on Leonardo da Vinci—the original working title was "Creative Production"—is an examination of this concept. See Adrian Parr, *Exploring the Work of Leonardo da Vinci within the Context of Contemporary Philosophical Thought and Art: From Bergson to Deleuze* (Lewinston: Edwin Mellen, 2003).

5. Hawken, *Blessed Unrest*, 18.

6. Michael Hardt and Antonio Negri, *Multitude: War and Democracy in the Age of Empire* (New York: Penguin, 2004).

Bibliography

Adelman, Larry, prod. "Racial Preferences for Whites: The Houses That Racism Built," *Race the Power of an Illusion*, PBS, California Newsreel, 2003.

Adorno, Theodor W. *Aesthetic Theory*, trans. Robert Hullot-Kentor (Minneapolis: University of Minnesota Press, 1998).

Agamben, Giorgio. *Homo Sacer: Sovereign Power and Bare Life*, trans. Daniel Heller-Roazen (Stanford, CA: Stanford University Press, 1998).

Albers, Manuel B. "The Double Function of the Gate: Social Inclusion and Exclusion in Gated Communities and Security Zones," Conference paper for *Gated Communities: Building Social Division or Safer Communities? 2003.*

Annan, Kofi. "International Strategy for Disaster Reduction," October 11, 2006. http://www.unisdr.org/eng/public_aware/world_camp/2006-2007/iddr/2006-iddr .htm.

Architecture for Humanity, ed. *Design Like You Give a Damn: Architectural Responses to Humanitarian Crises* (New York: Metropolis Books, 2006).

Architecture for Humanity. "Bay View Model Block Project." http://www.architec tureforhumanity.org/?q=node/165.

Bacevich, Andrew J. *The New American Militarism: How Americans Are Seduced by War* (Oxford: Oxford University Press, 2005).

Ballentine, Karen, and Jake Sherman, eds. *The Political Economy of Armed Conflict: Beyond Green and Grievance* (Boulder, CO: Lynne Rienner Publishers, 2003).

BAN. "Exporting Harm: The Hightech Trashing of Asia," *Press Release*, October 24, 2005.

BAN and Silicon Valley Toxics Coalition. "High Tech Toxic Trash from USA Found to Be Flooding Asia," *Press Release*, October 24, 2005.

Bang, Jan Martin. *Ecovillages: A Practical Guide to Sustainable Communities* (Gabriola Island, Canada: New Society Publishers, 2005).

Barnett, Thomas. *The Pentagon's New Map: War and Peace in the Twenty-First Century* (New York: Putnam, 2004).

Basel Convention on the Control of Transboundary Movements of Hazardous Wastes and Their Disposal, 2. http://www.basel.int/text/con-e-rev.pdf. Accessed June 6, 2008.

BBC News. "Q&A: The U.S. and Climate Change," February 11, 2002. http://news .bbc.co.uk/2/hi/americas/1820523.stm.

BBC News. "Huge Rise in Iraqi Death Tolls," October 11, 2006, http://news.bbc .co.uk/2/hi/middle_east/6040054.stm.

Bell, Daniel. *The Coming of Post-Industrial Society: A Venture into Social Forecasting* (New York: Basic Books, 1973).

Boggs, Carl. *Imperial Delusions: American Militarism and Endless War* (Landham, MD: Rowman & Littlefield, 2005).

Bótas, Paulo. "Rio de Janeiro to Spend US$1 Billion on Innovative Slum Improvement Programme," *City Mayors Environment*, December 5, 2003. http://www.city mayors.com/report/rio_favelas.html.

Broder, John M. "At Climate Meeting Bush Does Not Specify Goals," *New York Times*, September 29, 2007. http://www.nytimes.com/2007/09/29/washington/29climate .html?ref=americas.

Bromage, Arthur W., and John A. Perkins. "Willow Run Produces Bombers and Intergovernmental Problems," *The American Political Science Review* 36, no. 4 (August 1942), 689–697.

Brouwer, Joke, Philip Brookman, and Arjen Mulder. *Transurbanism* (Rotterdam, Netherlands: V2Publishing/NAi Publishing, 2002).

Browning, William D., Dianna Lopez Barnett, Mark Ginsberg, Anne Sprunt-Crawley, "Greening the White House: A Comprehensive Energy and Environmental Retrofit." http://www.rmi.org/images/PDFs/BuildingsLand/D94-09_GrnWhiteHouse.pdf.

Buchanan, Ian, and Gregg Lambert, eds. *Deleuze and Space* (Edinburgh: Edinburgh UP, 2005).

Buchanan, Ian. *Fredric Jameson: Live Theory* (London: Continuum, 2006).

Burkeman, Oliver. "The Warm-Mongers Rest Their Case," *The Age*, February 4, 2007.

Butler, Judith. *Bodies That Matter: On the Discursive Limits of Sex* (London: Routledge, 1993).

Butler, Judith. *Excitable Speech: A Politics of the Performative* (New York: Routledge, 1997).

Butler, Judith. "Vikki Bell: Interview with Judith Butler," *Theory, Culture & Society* 16, no. 2 (1999): 163–174.

Butler, Judith. *Kritik der ethischen Gewalt. Adorno Lectures, 2002* (Frankfurt: Institut für Sozialforschung an der Johann Wolfgang Goethe-Universität, 2003).

Calthorpe, Peter. *The Next American Metropolis: Ecology, Community, and the American Dream* (New York: Princeton Architectural Press, 1995).

Canizaro, Vincent B., ed. *Architectural Regionalism: Collected Writings on Place, Identity, Modernity, and Tradition* (New York: Princeton Architectural Press, 2007).

Care International, *Children and Poverty Campaign.* http://www.care.org/campaigns/childrenpoverty/index.asp?source=170740250000&WT.srch=1.

Carter, Jimmy. "Energy Policy," April 18, 1977. http://www.pbs.org/wgbh/amex/carter/filmmore/ps_energy.html.

Carter, Jimmy. "The Crisis of Confidence," July 15, 1979. http://www.pbs.org/wgbh/amex/carter/filmmore/ps_crisis.html.

Center for Media and Democracy. "Source Watch." http://www.sourcewatch.org/index.php?title=Global_Climate_Coalition.

Christian Aid. "The Climate of Poverty: Facts, Fears and Hopes," *Christian Aid Report,* May 2006. http://www.christianaid.org.uk/Images/climate_of_poverty_tcm15-21613.pdf.

Chung, Chuihua Judy, ed. *Harvard School of Design Guide to Shopping* (Hong Kong: Taschen, 2002).

Clapp, Jennifer. "Seeping through the Regulatory Cracks," *SAIS Review* xxii, no. 1 (Winter/Spring, 2002): 141–155.

Clinton, William J. *First Inaugural,* January 20, 1993. http://www.millercenter.virginia.edu/scripps/digitalarchive/speeches/clinton/wjc_1993_0120.

Clinton, Bill. "Reaffirming the U.S. Commitment to Protect Global Environment," April 26, 1993.

Corbett, Charles J., and Richard P. Turco. "Film and Television," *Southern California Environmental Report Card 2006* (Los Angeles: University of California, Los Angeles Institute of the Environment, 2006).

Correa, Charles. *Housing and Urbanization: Building Solutions for People and Cities* (London: Thames and Hudson, 2000).

Corsellis, Tom, and Antonella Vitale. *Transitional Settlement, Displaced Populations,* Oxfam (Cambridge: University of Cambridge, 2005).

Cubitt, Sean. *EcoMedia* (Amsterdam, New York: Rodopi Press, 2005).

Daniels, Klaus. *The Technology of Ecological Building* (Basel, Switzerland: Birkhäuser Verlag, 1995).

Davis, Mike. *Planet of the Slums* (London: Verso, 2006).

Ann Deehan, "Prevention of Alcohol-Related Crime: Operationalizing Situational and Environmental Strategies," *Crime Prevention and Community Safety: An International Journal* 6, no. 1 (2004): 43–51.

Deleuze, Gilles. *Nietzsche and Philosophy*, trans. Hugh Tomlinson (New York: Columbia University Press, 1983), 40.

Deleuze, Gilles. *Foucault*, trans. Seán Hand (Minneapolis: University of Minnesota Press, 1988).

Deleuze, Gilles. *Difference and Repetition*, trans. Paul Patton (New York: Columbia University Press, 1994).

Deleuze, Gilles. *Negotiations: 1972–1990* (New York: Columbia UP, 1995).

Deleuze, Gilles, and Félix Guattari. *Anti-Oedipus: Capitalism and Schizophrenia*, trans. Robert Hurley, Mark Seem, and Helen R. Lane (Minneapolis: University of Minnesota Press, 1977).

Deleuze, Gilles, and Félix Guattari. *A Thousand Plateaus: Capitalism and Schizophrenia*, trans. Brian Massumi (London: Athlone, 1987).

Deleuze, Gilles, and Félix Guattari. *What is Philosophy?* trans. Graham Burchell and Hugh Tomlinson (London: Verso, 1994).

De-Shalit, Avner. *The Environment: Between Theory and Practice* (Oxford: Oxford University Press, 2000).

Dorrian, Mark, and Gillian Rose, eds. *Deterritorializations . . . Revisioning Landscapes and Politics* (London: Black Dog Publishing, 2003).

De Soto, Hernando. *The Mystery of Capital: Why Capitalism Triumphs in the West and Fails Everywhere Else* (New York: Basic Books, 2003).

Douglas, Mary. *Purity and Danger: An Analysis of Concepts of Pollution and Taboo* (London: Routledge, 1966).

Dowling, Robyn. "Neotraditionalism in the Suburban Landscape: Cultural Geographies of Exclusion in Vancouver, Canada," *Urban Geography* 19, (1998): 105–122.

Durant, Robert F. *The Greening of the U.S. Military: Environmental Policy, National Security, and Organizational Change* (Washington, DC: Georgetown University Press, 2007).

Dwivedi, O. P., and Dhirendra K. Vajpeyi (ed.). *Environmental Policies in the Third World: A Comparative Analysis* (Westport, CT: Greenwood Press, 1995).

Ecorazzi. "President Bush Texas Ranch is Off-Grid, Eco-Friendly!" http://www
.ecorazzi.com/?p=1601.

Edwards, Andres R. *The Sustainability Revolution: Portrait of a Paradigm Shift* (Philadelphia: New Society Publishers, 2005).

Pelkonen, Eeva-Liisa, and Donald Albrecht, eds. *Eero Saarinen: Shaping the Future* (New Haven, CT: Yale University Press, 2006).

Eliot, John L. "A Dump Reviled, Revered," *National Geographic* 203, no. 1 (January 2003): 8–9.

Environmental Media Association. "17th Annual EMA Awards." http://www
.ema-online.org/.

Fiori, Jorge, Liz Riley, Ronaldo Ramirez. *Urban Poverty Alleviation through Environmental Upgrading in Rio de Janeiro: Favela Bairro*, Draft Research Report, R7343, March 2000.

Fiori, Jorge, Liz Riley, and Ronaldo Ramirez, "Favela Bairro and a New Generation of Housing Programmes for the Urban Poor," *Geoforum* 32, no. 4 (November 2001). http://www.sciencedirect.com.

Fishman, Charles. *The Wal-Mart Effect* (New York: Penguin, 2006).

Faler, Brian. "White House Turns Up Heat With Solar Energy at Spa," *Washington Post*, February 3, 2003.

Flannery, Tim. *The Weather Makers: How Man Is Changing the Climate and What It Means for Life on Earth* (New York: Atlantic Monthly Press, 2005).

Fleischer, Ari. "Statement by the Press Secretary on the Geneva Convention," White House, May 7, 2003. http://www.whitehouse.gov/news/releases/2003/05/20030507
-18.html.

Foote, Lee. "NASUSG Calls for Freeze on Polar Bear Reclassification," *Sustainable: The Newsletter of the IUCN SSC Sustainable Use Specialist Group*, June 2006.

Foster, Hal. *Design and Crime [And Other Diatribes]* (London: Verso, 2002).

Foster, John Bellamy. *Marx's Ecology: Materialism and Nature* (New York: Monthly Review Press, 2000).

Foucault, Michel. *Discipline and Punish: The Birth of the Prison*, trans. Alan Sheridan (London: Penguin Books, 1977).

Foucault, Michel. *The Archaeology of Knowledge and the Discourse on Language*, trans. A. M. Sheridan Smith (New York: Pantheon, 1982).

Foucault, Michel. *The History of Sexuality*, trans. William McNeill and Karen S. Feldman (Malden, MaA: Blackwell, 1998).

Frost, Barbara. "Water and Sanitation: The Silent Emergency," *UN Chronicle* xlv, no. 1, 2008. http://www.un.org/Pubs/chronicle/2007/issue4/0407p87.html. Accessed June 9, 2008.

Fuller, Henning, and Nadine Marquardt. "More Than a New Place to Live—A Whole New Way of Life," conference paper for *4th International Conference of the Research Network Private Urban Governance and Gated Communities, 2007.*

Global Green USA. "Global Green Promotes Plug-in Hybrid, All Electric, and Alternative Fuel Vehicles for Oscars Drive." http://www.globalgreen.org/press/releases/2007_2_23_oscarsgreencars.htm.

Goodman, Amy. "Exclusive: Facing Seven Years in Jail, Environmental Activist Daniel McGowan Speaks Out about the Earth Liberation Front, the Green Scare, and the Government's Treatment of Activists as 'Terrorists,'" *Democracy Now*, June 11, 2007. http://www.democracynow.org/article.pl?sid=07/06/11/142258.

Graham, Kathryn, and Wells, Samantha. "Aggression among Young Adults in the Social Context of the Bar," *Addiction Research* 9, 2001: 193–219.

Green Car Congress. "February 2006 US Hybrid Sales up 44% from Prior Year; Prius Down 7.5%," March 7, 2006. http://www.greencarcongress.com/2006/11/gms _design_conc.html

Green Car Congress. "GM's Design Concept HUMMER O$_2$: Fuel Cell HUMMER That Breathes," July 18, 2007. http://www.greencarcongress.com/2006/03/february_2006 _u.html

Greenwald, Noah D., and Kieran F. Suckling. "Progress or Extinction? A Systematic Review of the U.S. Fish and Wildlife Service's Endangered Species Act Listing Program 1974–2004," *Center for Biological Diversity*, May 2005.

Harasym, Sarah. *The Postcolonial Critic: Interviews, Strategies, Dialogues* (London: Routledge, 1990).

Hardt, Michael, and Antonio Negri. *Empire* (Cambridge, MA: Harvard University Press, 2000).

Hardt, Michael, and Antonio Negri. *Multitude: War and Democracy in the Age of Empire* (New York: Penguin, 2004).

Harris, Paul G. *The Environment, U.S. International Relations, and Foreign Policy* (Washington, DC: Georgetown University Press, 2001).

Harris, Richard. "Climate Roadmap Emerges from Grueling Bali Talks," *National Public Radio*, December 15, 2007. http://www.npr.org/templates/story/story .php?storyId=17519540.

Hawken, Paul. *The Ecology of Commerce: A Declaration of Sustainability* (New York: Harper Collins, 1993).

Hawken, Paul, Amory Lovins, and L. Hunter Lovins. *Natural Capitalism: Creating the Next Industrial Revolution* (Boston: Little, Brown and Company, 1999).

Hawken, Paul. *Blessed Unrest: How the Largest Movement in the World Came into Being and Why No One Saw It Coming* (New York: Viking, 2007).

Higgs, Robert. "The Trillion Dollar Defense Budget Is Already Here," *The Independent Institute*, March 15, 2007. http://www.independent.org/newsroom/article.asp?id =1941.

Hozic, Aida. *Hollyworld: Space, Power, and Fantasy in the American Economy* (Ithaca: Cornell University Press, 2001).

Hunter, James Davison. *Culture Wars: The Struggle to Define America* (New York: Basic Books, 1991).

Huntington, Samuel P. *The Soldier and the State: The Theory and Politics of Civil-Military Relations* (Cambridge, MA: Belknap Press, 1981).

Ingram, David. *Green Screen: Environmentalism and Hollywood Cinema* (Exeter, Devon: University of Exeter Press, 2000).

Inuit Circumpolar Council (Canada). "Inuit Oppose and Seek Clarification of IUCN Decision to Change Polar Bear Status to 'Vulnerable,'" ICC press release, May 11, 2006.

Inuit Circumpolar Council (Canada). "Inuit Cite IPCC Results as Further Proof of Human Impacts Contributing to Climate Change: Inuit Call on Canada to Recognize Arctic in Foreign Policy and Commit Resources towards Adaptation," ICC press release, February 2, 2007.

Intergovernmental Panel on Climate Change Fourth Assessment Report Working Panel. *Climate Change 2007: Synthesis Report* (IPCC).

Jasanoff, Sheila, and Marybeth Long Martello, eds. *Earthly Politics: Local and Global in Environmental Governance* (Cambridge, MA: MIT Press, 2004).

Jameson, Fredric. "Metacommentary," *PMLA* 86, no. 1 (January 1971): 9–18.

Jameson, Fredric. *The Political Unconscious* (Ithaca, NY: Cornell University Press, 1981).

Jameson, Fredric. "Third-World Literature in the Era of Multinational Capitalism," *Social Text* 15 (1986): 65–88.

Jameson, Fredric. *Signatures of the Visible* (London: Routledge, 1992).

Jameson, Fredric. *Postmodernism: Or the Cultural Logic of Late Capitalism* (Durham: Duke University Press, 1999).

Jameson, Fredric. "Future City," *New Left Review* 21, (May/June 2003): 65–79.

Jellicoe, Geoffrey, and Susan Jellicoe. *The Landscape of Man: Shaping the Environment from Prehistory to the Present Day* (New York: Thames & Hudson, 1995).

Jones, David Lloyd. *Architecture and the Environment* (Woodstock, NY: Overlook Press, 1998).

Kaika, Maria. *City of Flows: Modernity, Nature, and the City* (New York: Taylor and Francis, 2005).

Kapferer, Jean-Noël. *Strategic Brand Management: New Approaches to Creating and Evaluating Brand Equity* (New York: Free Press, 1992).

Keller, Kevin L. *Strategic Brand Management* (Upper Saddle River, NJ: Prentice Hall, 1998).

Klein, Naomi. *The Shock Doctrine: The Rise of Disaster Capitalism* (New York: Metropolitan Books, 2007).

Kristeva, Julia. *Powers of Horror: An Essay on Abjection*, trans. Leon S. Roudiez (New York: Columbia UP, 1982).

Kumar, Naresh. "A Spatial Dependency and Causality Analysis of Crime in Savannah, Georgia, 2000," *Department of Geography Papers*, University of Iowa, 2005.

Lake, Anthony. "From Containment to Enlargement," delivered at Johns Hopkins University on September 21, 1993. http://www.mtholyoke.edu/acad/intrel/lakedoc .html.

LePla, Joseph, Susan Voeller Davis, and Lynne M. Parker. *Brand Driven: The Route to Integrated Branding through Great Leadership* (London: Kogan Page, 2003).

Levitt, Theodore. *The Marketing Mode* (New York: McGraw-Hill, 1969).

Lindsey, Brink. *The Age of Abundance: How Prosperity Transformed America's Politics and Culture* (New York: Harper Collins, 2007).

Local Authority Pension Fund Forum. "Local Authorities Oppose BP Pay Policy's Lack of Safety Targets," press release, March 26, 2007.

Lockwood, Daniel. "Mapping Crime in Savannah," *Social Science Computer Review* 25, no. 2 (2007): 194–209.

Lutter, Randall, and Jason F. Shogren, eds. *Painting the White House Green: Rationalizing Environmental Policy inside the Executive Office of the President* (Washington, DC: Resources for the Future Press, 2004).

Lütticken, Sven. "Unnatural History," *New Left Review* 45, (May/June 2007): 115–131.

Luymes, Don. "The Fortification of Suburbia: Investigating the Rise of Enclave Communities," *Landscape and Urban Planning*, no. 39 (1997): 187–203.

Machado, Rodolfo. "Memoir of a Visit," in ed. Rodolfo Machado, *The Favela-Bairro Project* (Cambridge, MA: Harvard Design School, 2003).

Massachusetts Institute of Technology. "Tsunami—Safe(R) House: A Design for the Prajnopaya Foundation." http://senseable.mit.edu/tsunami-prajnopaya.

May, Lary. *The Big Tomorrow: Hollywood and the Politics of the American Way* (Chicago: Chicago University Press, 2000).

McCarthy, Joseph. "Speech on Communists in the State Department," 1950. http://www.civics-online.org/library/formatted/texts/mccarthy.html.

McDonough, William. *The Hannover Principles* (New York: 1992).

McDonough, William, and Michael Braungart. *Crade-to-Cradle: Remaking the Way We Make Things* (New York: Northpoint Press, 2002).

McHarg, Ian. *Design with Nature* (New York: John Wiley & Sons, 1995).

McIntosh, William Rebecca Murray, John Murray, and Debra Sabia, "Are the Liberal Good in Hollywood? Characteristics of Political Figures in Popular Films from 1945 to 1998," *Communications Reports* 16 (2003): 57–68.

Mehrotra, Rahul, ed. *Everyday Urbanism*, Michigan Debates on Urbanism, vol. 1 (Ann Arbor, MI: University of Michigan, 2005).

Mock, Gregory. "How Much Do We Consume?" *World Resources 2000–2001*, June 2000. http://earthtrends.wri.org/features/view_feature.php?theme=6&fid=7.

Monahan, Torin. "Electronic Fortification in Phoenix," *Urban Affairs Review* 42, no. 2 (2006): 169–192.

Monbiot, George. *Heat: How to Stop the Planet from Burning* (Cambridge, MA: South End Press, 2007).

Monbiot, George. "Environmental Feedback: A Reply to Clive Hamilton," *New Left Review*, no. 45 (May/June 2007): 105–113.

National Resource Defense Council. "Environmental Achievements of the 79th Annual Academy Awards." http://www.nrdc.org.

Nelson, Cary, and Lawrence Grossberg, eds. *Marxism and the Interpretations of Culture* (London: Macmillan, 1988).

Neuwirth, Robert. *Shadow Cities: A Billion Squatters, a New Urban World* (New York: Routledge, 2006).

Nietzsche, Friedrich. *The Will to Power*, trans. Walter Kaufmann and R. J. Hollingdale (New York: Vintage, 1968).

Nietzsche, Friedrich. *Human, All Too Human: A Book for Free Spirits*, trans. R. J. Hollingdale (Cambridge: Cambridge University Press, 1996).

Ohio Environmental Protection Agency (EPA). "2004 Out-of-State Waste," *Ohio EPA Fact Sheet*, November 2005.

Olgay, Victor, and Aladar Olgay. *Design with Climate: Bioclimatic Approach to Architectural Regionalism* (Princeton: Princeton University Press, 1963).

Oliver, Paul. *Built to Meet Needs: Cultural Issues in Vernacular Architecture* (New York: Architectural Press, 2006).

Oliver, Paul. *Dwellings: The Vernacular House Worldwide* (London: Phaidon, 2007).

One Laptop Per Child. http://laptop.org/vision/index.shtml.

Orr, David. W. *The Nature of Design: Ecology, Culture, and Human Intention* (Oxford: Oxford University Press, 2002).

Orr, David W. *The Last Refuge: Patriotism, Politics, and the Environment in an Age of Terror* (Washington: Island Press, 2005).

OSHA. "National News Release: USDL 05-1740," *National News Release*, September 22, 2005. http://www.osha.gov/pls/oshaweb/owadisp.show_document?p_table =NEWS_RELEASES&p_id=11589.

Oxfam International. "The Tsunami's Impact on Women," *Oxfam Briefing Note*, March 2005. http://www.oxfam.org.uk/what_we_do/issues/conflict_disasters/downloads/bn_tsunami_women.pdf.

Parr, Adrian. *Exploring the Work of Leonardo da Vinci within the Context of Contemporary Philosophical Thought and Art: From Bergson to Deleuze* (Lewinston: Edwin Mellen, 2003).

Parr, Adrian. *Memorial Culture: Desire, Singular Memory, and the Politics of Trauma* (Edinburgh: Edinburgh University Press, 2008).

Patton, Paul. *Deleuze and the Political* (London: Routledge, 2000).

Payne, Geoffrey, and Michael Majale. *The Urban Housing Manual: Making Regulatory Frameworks for the Poor* (London: Earthscan, 2004).

Perlman, Janice. E. "The Myth of Marginality Revisited: The Case of *Favelas* in Rio de Janeiro, 1969–2003," March 24, 2005.

Perlman, Janice. E. *The Myth of Marginality: Urban Poverty and Politics in Rio de Janeiro* (Berkeley: University of California Press, 1980).

Perlman, Janice. E. *Longitudinal Panel Studies in Squatter Communities: Lessons from a re-study of Rio's favelas: 1969–2003*, Joint World Bank Workshop Urban Longitudinal Research Methodology. London, May 28–29, 2003.

Peterson, Tarla Rai, ed. *Green Talk in the White House: The Rhetorical Presidency Encounters Ecology* (College Station: Texas A&M University, 2004).

"The PlayPump Water System," *PlayPumps International*. http://www.playpumps
.org/site/c.hqLNIXOEKrF/b.2589561/k.C08/The_PlayPump_System__The_Water
_Problem.htm. Accessed June 9, 2008.

Postman, Neil. *Amusing Ourselves to Death: Public Discourse in the Age of Show Business*
(New York: Penguin, 1986), http://www.worldbank.org/urban/symposium2005/
papers/perlman.pdf.

Rancière, Jacques. *The Politics of Aesthetics*, trans. Gabriel Rockhill (London: Continuum, 2004).

Ranciére, Jacques. *Hatred of Democracy*, trans. Steve Concoran (London: Verso,
2006).

Reis, Rejane. "Teaching Tourism in the Slums," *Favela Rocinha Tourism Workshop*.
http://www.favelatourismworkshop.com.

Rogers, Heather. *Gone Tomorrow: The Hidden Life of Garbage* (New York: New Press,
2005).

Ronald Reagan Obituary. *New York Times*, June 7, 2004. http://query.nytimes.com/
gst/fullpage.html?res=9D01E4DE1E31F934A35755C0A9629C8B63.

Rosen, Nick. "Meanwhile, Back at the Ranch," *Off-Grid*, February 18, 2007. http://
www.ecorazzi.com/?p=1601.

Rosenblatt, Roger, and San Bruno. "The Man Who Wants Buildings to Love Kids,"
Time Magazine, February 15, 1999, http://www.time.com/time/reports/environ
ment/heroes/heroesgallery/0,2967,mcdonough2,00.html.

Rosenzweig, Michael L., *Win-Win Ecology: How the Earth's Species Can Survive in the
Midst of Human Enterprise* (Oxford: Oxford University Press, 2003).

Roy, Arundhati. *The Algebra of Infinite Justice* (London: Flamingo, 2002).

Sachs, Jeffrey. *The End of Poverty: Economic Possibilities for Our Time* (New York: Penguin, 2005).

Said, Edward. *Orientalism* (New York: Vintage, 1978).

Sasi, K. P., and Max Martin "Shelter: Little Problems Need Big Attention," *Tsunami
Response Watch.org*. http://www.tsunamiresponsewatch.org/2006/11/02/shelter-little
-problems-need-big-attention/#more-724.

Savannah Chatham Metropolitan Police. "2006 Neighborhood Crime Statistics,"
2006.

Savannah Live Well and Prosper Data. http://www.seda.org/content.php?section
=data&subsection=population.

Scribner, J., D. MacKinnon, and J. Dwyer. "The Risk of Assaultive Violence and Alcohol Availability in Los Angeles County," *American Journal of Public Health* 85, no. 3 (1995): 335–340.

Scott, Lee. "Twenty-First Century Leadership," October 24, 2005. http://www.walmartstores.com/Files/21st%20Century%20Leadership.pdf.

Scully, Malcolm G. "Making Peace with Diversity," *Chronicle of Higher Education* 49, no. 29: B15.

Shelton, Ted. "Greening the White House: Executive Mansion as Symbol of Sustainability," *Journal of Architectural Education* 60, no. 4 (May 2007): 31–38.

Shiva, Vandana. *Biopiracy: The Plunder of Nature and Knowledge* (Cambridge, MA: South End Press, 1997).

Soares, Fabio, and Yuri Soares. *The Socio-Economic Impact of Favela-Bairro: What Do the Data Say?* Inter-American Development Bank, Washington, DC, August 2005.

Social Investment Forum. *2005 Socially Responsible Investing Trends in the United States*, February 24, 2005. Washington DC.

Sorkin, Michael, ed. *Variations on a Theme Park: The New American City and the End of Public Space* (New York: Noonday Press, 1992).

Spinoza, Baruch. *Ethics*, trans. G. H. R. Parkinson (Oxford: Oxford University Press, 2000).

Stecklow, Steve. "A Little Laptop with Big Ambitions," *Wall Street Journal*, November 24, 2007: A1.

Stockholm International Peace Research Institute (SIPRI). *SIPRI Yearbook 2007: Armaments, Disarmament, and International Security* (Oxford: Oxford University Press, 2007).

Strasser, Susan. *Waste and Want: A Social History of Trash* (New York: Holt, 2000).

Strickland, Roy, ed. *Post Urbanism and ReUrbanism* (Ann Arbor, MI: University of Michigan, 2005).

Stupak, Bart. "Stupak BP Hearing Statement," *News from Congressman Bart Stupak*, September 7, 2006, http://www.house.gov/list/press/mi01_stupak/090706BPHearing Statement.html

Target Opportunity: Earth Liberation Front. http://www.targetofopportunity.com/elf.htm.

Thakker, Nisha. "India's Toxic Landfills: A dumping Ground for the World's Electronic Waste," *Sustainable Development Law and Policy*, (Spring 2006): 58–80.

Till, Karen. "Neotraditional Towns and Urban Villages: The Cultural Production of a Geography of Otherness," *Environment and Planning D: Society and Space* 11, (1993): 709–732.

United Nations. "Press Conference by United Nations Deputy Emergency Relief Coordinator on Recent Floods in South Asia," August 9, 2007. http://www.un.org/News/briefings/docs/2007/070809_Wahlstrom.doc.htm.

UN-HABITAT. *The Challenge of Slums: Global Report on Human Settlements 2003* (London: Earthscan, 2003).

UN-HABITAT. *Financing Urban Shelter: Global Report on Human Settlements 2005* (London: Earthscan, 2005).

UN-HABITAT. *State of the World's Cities, 2006–2007: The Millennium Development Goals and Urban Sustainability* (London: Earthscan, 2006).

The United States Commission on National Security/21st Century. *Road Map for National Security: Imperative for Change*, February 15, 2001.

U.S. Army. "U.S. Army Sustainability—Goals." http://www.sustainability.army.mil/overview/goals.cfm. Accessed June 29, 2007.

U.S. Census Bureau. "Poverty." http://www.census.gov/hhes/www/poverty/poverty 06/pov06hi.html.

U.S. Department of Defense. *FY 2007 Department of Defense Budget*, February 6, 2006, http://www.defenselink.mil/news/Feb2006/d20060206slides.pdf.

U.S. Department of Energy. "State Energy Alternatives, U.S. Department of Energy: Energy Efficiency and Renewable Energy." http://www.eere.energy.gov/states/alternatives/government_buildings.cfm.

U.S. Environmental Protection Agency. "Recycling." http://www.epa.gov/epaoswer/non-hw/muncpl/recycle.htm.

Vaill, Sarah. *The Calm In the Storm: Women Leaders in Gulf Coast Recovery* (Women's Funding Network, 2006).

Wagoner, Rick. "2005/2006 Corporate Responsibility Report," *GM Corporate Responsibility.* http://www.gm.com/corporate/responsibility.

Walker, Liz. *Ecovillage at Ithaca: Pioneering a Sustainable Culture* (Gabriola Island, Canada: New Society Publishers, 2005).

Wal-Mart Fact Sheet. "Wal-Mart Opens First Experimental Supercenter," April 10, 2007. http://www.walmartfacts.com/articles/2081.aspx.

Wal-Mart Realty. "Basic List of Available Buildings." http://wal-martrealty.com/Buildings/PrintableBuilding/BasicBldgListOnly.html.

Warner, Melanie. "Wal-Mart Eyes Organic Food," *New York Times*, May 12, 2006.

White House Historical Association. http://www.whitehousehistory.org/.

White House. *The Greening of the White House: Six Year Report*, November 1999.

Williams, Erica, Olga Sorokina, Avis Jones-DeWeever, and Heidi Hartmann. "The Women of New Orleans and the Gulf Coast: Multiple Disadvantages of Key Assets for Recovery, Part II. Gender, Race, and Class in the Labor Market," *Briefing Paper*, Institute for Women's Policy Research, August 2006.

World Bank. "Tracing the Evolution of Poverty in Rio's Favelas." http://lnweb18 .worldbank.org/LAC/LAC.nsf/ECADocbyUnid/C9413A90E66702C085256F9A004B7 9F8?Opendocument.

World Conservation Union. "2006 IUCN Red List of Threatened Species: Media Package for North America," 2006. http://www.iucn.org.

Worldwatch Institute. *State of the World 2007: Our Urban Future* (New York: W.W. Norton & Company, 2007).

Wright, Bruce. "Poppin' Fresh Retail," *Print (New York)* 61, no. 4 (July/August 2007): 62–67.

Yeang, Ken. *Designing with Nature: the Ecological Basis for Architectural Design* (New York: McGraw Hill, 1995).

Young, William. "A Dump No More," *American Forests* 101, (August 1995): 58–9.

Yunus, Muhammad. *Creating a World without Poverty* (New York: Public Affairs, 2007).

Zeiher, Laura. *The Ecology of Architecture: The Complete Guide to Creating the Environmentally Conscious Building* (New York: Whitney Library of Design, 1996).

Zhu, L., D. M. Gorman, and S. Horel. "Alcohol Outlet Density and Violence: A Geospatial Analysis," *Alcohol and Alcoholism* 39, no. 4 (2004): 369–375.

Žižek, Slavoj. *The Parallax View* (Cambridge, MA: MIT Press, 2006).

Index

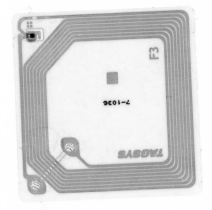